NANOTECHNOLOGY SCIENCE AND TECHNOLOGY

A FIRST COURSE ON BASIC ELEMENTS OF HEAT FLOW IN NANOPOROUS FABRICS

NANOTECHNOLOGY SCIENCE AND TECHNOLOGY

Additional books in this series can be found on Nova's website under the Series tab.

Additional E-books in this series can be found on Nova's website under the E-book tab.

NANOTECHNOLOGY SCIENCE AND TECHNOLOGY

A FIRST COURSE ON BASIC ELEMENTS OF HEAT FLOW IN NANOPOROUS FABRICS

A. K. HAGHI

Nova Science Publishers, Inc.
New York

Copyright © 2012 by Nova Science Publishers, Inc.

All rights reserved. No part of this book may be reproduced, stored in a retrieval system or transmitted in any form or by any means: electronic, electrostatic, magnetic, tape, mechanical photocopying, recording or otherwise without the written permission of the Publisher.

For permission to use material from this book please contact us:
Telephone 631-231-7269; Fax 631-231-8175
Web Site: http://www.novapublishers.com

NOTICE TO THE READER

The Publisher has taken reasonable care in the preparation of this book, but makes no expressed or implied warranty of any kind and assumes no responsibility for any errors or omissions. No liability is assumed for incidental or consequential damages in connection with or arising out of information contained in this book. The Publisher shall not be liable for any special, consequential, or exemplary damages resulting, in whole or in part, from the readers' use of, or reliance upon, this material. Any parts of this book based on government reports are so indicated and copyright is claimed for those parts to the extent applicable to compilations of such works.

Independent verification should be sought for any data, advice or recommendations contained in this book. In addition, no responsibility is assumed by the publisher for any injury and/or damage to persons or property arising from any methods, products, instructions, ideas or otherwise contained in this publication.

This publication is designed to provide accurate and authoritative information with regard to the subject matter covered herein. It is sold with the clear understanding that the Publisher is not engaged in rendering legal or any other professional services. If legal or any other expert assistance is required, the services of a competent person should be sought. FROM A DECLARATION OF PARTICIPANTS JOINTLY ADOPTED BY A COMMITTEE OF THE AMERICAN BAR ASSOCIATION AND A COMMITTEE OF PUBLISHERS.

Additional color graphics may be available in the e-book version of this book.

LIBRARY OF CONGRESS CATALOGING-IN-PUBLICATION DATA

A first course on basic elements of heat flow in nanoporous fabrics / editor, A. K. Haghi.
    p. cm.
  Includes bibliographical references and index.
  ISBN 978-1-61942-938-3 (soft cover)
  1. Porous materials. 2. Nanostructured materials. I. Haghi, A. K.
  TA418.9.P6F57 2011
  620.1'16--dc23
                    2012000891

Published by Nova Science Publishers, Inc. † New York

CONTENTS

Preface		vii
About the Author		ix
Nomenclature		1
Chapter 1	Basic Concepts	5
Chapter 2	Convective Heat Flow	19
Chapter 3	Conductive Heat Flow	49
Chapter 4	Radiative Heat Flow	61
Chapter 5	Heat Flow and Clothing Comfort	85
Appendix		117
Index		153

PREFACE

The textile industry views fibers as natural or synthetic filament, such as cotton or nylon, capable of being spun into yam, or simply as material made of such filaments.

The discovery and rapid evolution of nanofibers have led to a vastly improved understanding of nanotechnology, as well as dozens of possible applications for nanomaterials of different shapes and sizes ranging from composites to biology, medicine, energy, transportation, and electronic devices.

For Heat flow analysis of wet porous nanostructure fabrics, the liquid is water and the gas is air. Evaporation or condensation occurs at the interface between the water and air so that the air is mixed with water vapor. A flow of the mixture of air and vapor may be caused by external forces, for instance, by an imposed pressure difference. The vapor will also move relative to the gas by diffusion from regions where the partial pressure of the vapor is higher to those where it is lower.

Heat flow in porous nanostructure fabrics is the study of energy movement in the form of heat which occurs in many types of processes. The transfer of heat in porous nanostructure fabrics occurs from the high to the low temperature regions. Therefore a temperature gradient has to exist between the two regions for heat transfer to happen. It can be done by conduction (within one porous solid or between two porous solids in contact), by convection (between two fluids or a fluid and a porous solid in direct contact with the fluid), by radiation (transmission by electromagnetic waves through space) or by combination of the above three methods.

In this book, the basic elements of heat flow in nano-porous textile materials is presented. The book provides polymer scientists and engineers with a comprehensive, practical "how-to" reference.

This book is intended to be essentially a practical guide; a detailed discussion of the theoretical side of the subject should be outside its scope. A considerable literature exists and the reader who wishes to make a further study of this aspect of the subject is referred to the bibliography at the end of this book. However, for a proper appreciation of what is involved to heat flow in porous nanostructure fabrics, certain basic theories and concepts should be understood.

ABOUT THE AUTHOR

Dr. A. K. Haghi holds a BSc in urban and environmental engineering from the University of North Carolina (USA), a MSc in mechanical engineering from North Carolina A&T State University (USA), a DEA in applied mechanics, acoustics, and materials from the Université de Technologie de Compiègne (France), and a PhD in engineering sciences from the Université de Franche-Comté (France).

He has written about 1000 original articles, 250 monographs, and 170 chapters in 40 volumes. It is apparent from this work that he has made valuable contributions to the theory and practice of Chemical Engineering, Heat and Mass Transfer, Porous Media, Industrial Drying, Polymers, Nanofibers and Nanocomposites.

Dr Haghi is Editor-In-Chief of International Journal of Chemoinformatics and Chemical Engineering and Editor-In-Chief of Polymers Research Journal. He is member of many editorial boards of journals published in U.S.A. He is Senior Editor of Apple Academic Press (Canada).

He served as associate member of University of Ottawa and was a member of Canadian Society of Mechanical Engineering. He serves as faculty member at University of Guilan (Iran).

NOMENCLATURE

A	Area
a,b,c	Constants
c_p	constant pressure specific heat
C_A	moisture content of air in fabric pores
C_a	water-vapor concentration in the air filling the inter-fiber void space
C_e	moisture content of extent air
C_F	moisture content of fibers in a fabric
C_f	water-vapor concentration in the fibers of the fabric (kg m^{-3})
C_p	specific heat
D	diffusion coefficient
D_b	bound water conductivity
D_{eff}	effective diffusivity
E_d	activation energy of movement of bound water
H	enthalpy (J/kg)
h_e	Heat transfer coefficient
h_m	mass transfer coefficient
Δh_v	enthalpy of vaporization (J/kg)
Δh_{vap}	latent heat of evaporation
J	species diffusion flux

J_L		free water flux
K		Permeability
K_0		single phase permeability of porous material
K_r		relative permeability
k		thermal conductivity
k_{eff}		effective thermal conductivity
m		Ratio of diffusion coefficients of air and water vapour
m'		mass source per unit volume
$\dot{m}$		evaporation rate, mass transfer rate
M		molecular weight
P		pressure
P_c		capillary pressure
P_s		saturation pressure
q		convective heat transfer rate
Q		entholpy of desorption from solid phase
R		Radius
R		gas constant, Fiber regain
S		Pore saturation
T		Time
T		Temperature
T_e		external air temperature
U		moisture content

Greek Symbols

γ		Pore volume density function
λ		latent heat of evaporation
λ_{eff}		effective thermal conductivity
μ		Viscosity
v		Fluid velocity
ρ		Density
σ		surface tension

τ	tortuosity factor of capillary paths
ψ	relative humidity
ω	averaging volume
ε	volume fraction (m^3 of quantity $/m^3$)

Subscripts

0	Initial
C	capillary, critical
Eq	Equilibrium
G	Gas
Ir	Irreducible
Ms	maximum sorptive
V	Vapour
W	Water
β	Liquid phase
γ	Gas phase
σ	Solid phase
Bl	Bound liquid
Ds	Dry solid
Lv	Liquid-to-vapor
Ls	Liquid-to-solid
Sat	Saturation
Sv	Solid-to-vapor
V	Vapor

Superscripts

G	intrinsic average over the gaseous phase
L	intrinsic average over the liquid phase
*	vapour saturated
-	average value

Chapter 1

BASIC CONCEPTS

1.1. INTRODUCTION

Heat and mass transfer in wet porous media are coupled in a complicated way. The structure of the solid matrix varies widely in shape. There is, in general, a distribution of void sizes, and the structures may also be locally irregular. Energy transport in such a medium occurs by conduction in all of the phases. Mass transport occurs within voids of the medium. In an unsaturated state these voids are partially filled with a liquid, whereas the rest of the voids contain some gas. It is a common misapprehension that non-hygroscopic fibers (i.e., those of low intrinsic for moisture vapor) will automatically produce a hydrophobic fabric. The major significance of the fine geometry of a textile structure in contributing to resistance to water penetration can be stated in the following manner:

The requirements of a water repellent fabric are (*a*) that the fibers shall be spaced uniformly and as far apart as possible and (*b*) that they should be held so as to prevent their ends drawing together. In the meantime, wetting takes place more readily on surfaces of high fiber density and in a fabric where there are regions of high fiber density such as yarns, the peripheries of the yarns will be the first areas to wet out and when the peripheries are wetted, water can pass unhindered through the fabric.

For thermal analysis of wet fabrics, the liquid is water and the gas is air. Evaporation or condensation occurs at the interface between the water and air so that the air is mixed with water vapor. A flow of the mixture of air and vapor may be caused by external forces, for instance, by an imposed pressure difference. The vapor will also move relative to the gas by diffusion from regions where the partial pressure of the vapor is higher to those where it is lower.

Again, heat transfer by conduction, convection, and radiation and moisture transfer by vapor diffusion are the most important mechanisms in very cool or warm environments from the skin.

Meanwhile, Textile manufacturing involves a crucial energy-intensive drying stage at the end of the process to remove moisture left from dye setting. Determining drying characteristics for textiles, such as temperature levels, transition times, total drying times and evaporation rates, etc is vitally important so as to optimize the drying stage. In general, drying means to make free or relatively free from a liquid. We define it more narrowly as the vaporization and removal of water from textiles.

1.2. HEAT

When a wet fabric is subjected to thermal drying two processes occur simultaneously, namely:

a) Transfer of heat to raise the wet fabric temperature and to evaporate the moisture content.
b) Transfer of mass in the form of internal moisture to the surface of the fabric and its subsequent evaporation.

The rate at which drying is accomplished is governed by the rate at which these two processes proceed. Heat is a form of energy that can across the boundary of a system. Heat can, therefore, be defined as "the form of energy that is transferred between a system and its surroundings as a result of a temperature difference". There can only be a transfer of energy across the boundary in the form of heat if there is a temperature difference between the system and its surroundings. Conversely, if the system and surroundings are at the same temperature there is no heat transfer across the boundary.

Strictly speaking, the term "heat" is a name given to the particular form of energy crossing the boundary. However, heat is more usually referred to in thermodynamics through the term "heat transfer", which is consistent with the ability of heat to raise or lower the energy within a system.

There are three modes of heat transfer:

- Convection
- Conduction
- Radiation

All three are different. Convection relies on movement of a fluid. Conduction relies on transfer of energy between molecules within a solid or fluid. Radiation is a form of electromagnetic energy transmission and is independent of any substance between the emitter and receiver of such energy. However, all three modes of heat transfer rely on a temperature difference for the transfer of energy to take place.

The greater the temperature difference the more rapidly will the heat be transferred. Conversely, the lower the temperature difference, the slower will be the rate at which heat is transferred. When discussing the modes of heat transfer it is the rate of heat transfer Q that defines the characteristics rather than the quantity of heat.

As it was mentioned earlier, there are three modes of heat transfer, convection, conduction and radiation. Although two, or even all three, modes of heat transfer may be combined in any particular thermodynamic situation, the three are quite different and will be introduced separately.

The coupled heat and liquid moisture transport of nano-porous material has wide industrial applications in textile engineering and functional design of apparel products. Heat transfer mechanisms in nano-porous textiles include conduction by the solid material of fibers, conduction by intervening air, radiation, and convection. Meanwhile, liquid and moisture transfer mechanisms include vapor diffusion in the void space and moisture sorption by the fiber, evaporation, and capillary effects. Water vapor moves through textiles as a result of water vapor concentration differences. Fibers absorb water vapor due to their internal chemical compositions and structures. The flow of liquid moisture through the textiles is caused by fiber-liquid molecular attraction at the surface of fiber materials, which is determined mainly by surface tension and effective capillary pore distribution and pathways. Evaporation and/or condensation take place, depending on the temperature and moisture distributions. The heat transfer process is coupled with the moisture transfer processes with phase changes such as moisture sorption and evaporation.

Mass transfer in the drying of a wet fabric will depend on two mechanisms: movement of moisture within the fabric which will be a function of the internal physical nature of the solid and its moisture content; and the movement of water vapor from the material surface as a result of water vapor from the material surface as a result of external conditions of temperature, air humidity and flow, area of exposed surface and supernatant pressure.

1.3. CONVECTION HEAT TRANSFER

A very common method of removing water from textiles is convective drying. Convection is a mode of heat transfer that takes place as a result of motion within a fluid. If the fluid, starts at a constant temperature and the surface is suddenly increased in temperature to above that of the fluid, there will be convective heat transfer from the surface to the fluid as a result of the temperature difference. Under these conditions the temperature difference causing the heat transfer can be defined as:

ΔT = (surface temperature) − (mean fluid temperature)

Using this definition of the temperature difference, the rate of heat transfer due to convection can be evaluated using Newton's law of cooling:

$$Q = h_c A \Delta T \qquad (1.1)$$

where A is the heat transfer surface area and h_c is the coefficient of heat transfer from the surface to the fluid, referred to as the "convective heat transfer coefficient".

The units of the convective heat transfer coefficient can be determined from the units of other variables:

$$Q = h_c A \Delta T$$
$$W = (h_c) m^2 K$$

So the units of h_c are $W/m^2 K$.

The relationship given in equation (1.1) is also true for the situation where a surface is being heated due to the fluid having higher temperature than the surface. However, in this case the direction of heat transfer is from the fluid to the surface and the temperature difference will now be

Basic Concepts

ΔT = (Mean fluid temperature) - (Surface temperature)

The relative temperatures of the surface and fluid determine the direction of heat transfer and the rate at which heat transfer take place.

As given in equation (1.1), the rate of heat transfer is not only determined by the temperature difference but also by the convective heat transfer coefficient h_c. This is not a constant but varies quite widely depending on the properties of the fluid and the behavior of the flow. The value of h_c must depend on the thermal capacity of the fluid particle considered, i.e. mC_p for the particle. The higher the density and C_p of the fluid the better the convective heat transfer.

Two common heat transfer fluids are air and water, due to their widespread availability. Water is approximately 800 times more dense than air and also has a higher value of C_p. If the argument given above is valid then water has a higher thermal capacity than air and should have a better convective heat transfer performance. This is borne out in practice because typical values of convective heat transfer coefficients are as follows:

Fluid	$h_c (W/m^2 K)$
water	500-10000
air	5-100

The variation in the values reflects the variation in the behavior of the flow, particularly the flow velocity, with the higher values of h_c resulting from higher flow velocities over the surface.

1.4. CONDUCTION HEAT TRANSFER

If a fluid could be kept stationary there would be no convection taking place. However, it would still be possible to transfer heat by means of conduction. Conduction depends on the transfer of energy from one molecule to another within the heat transfer medium and, in this sense, thermal conduction is analogous to electrical conduction.

Conduction can occur within both solids and fluids. The rate of heat transfer depends on a physical property of the particular solid of fluid, termed its thermal conductivity k, and the temperature gradient across the medium. The thermal conductivity is defined as the measure of the rate of heat transfer across a unit width of material, for a unit cross-sectional area and for a unit difference in temperature.

From the definition of thermal conductivity k it can be shown that the rate of heat transfer is given by the relationship:

$$Q = \frac{kA\Delta T}{x} \qquad (1.2)$$

ΔT is the temperature difference $T_1 - T_2$, defined by the temperature on the either side of the porous surface. The units of thermal conductivity can be determined from the units of the other variables:

$$Q = kA\Delta T / x$$
$$W = (k)m^2 K / m$$

The unit of k is $W/m^2 K/m$.

1.5. RADIATION HEAT TRANSFER

The third mode of heat transfer, radiation, does not depend on any medium for its transmission. In fact, it takes place most freely when there is a perfect vacuum between the emitter and the receiver of such energy. This is proved daily by the transfer of energy from the sun to the earth across the intervening space.

Radiation is a form of electromagnetic energy transmission and takes place between all matters providing that it is at a temperature above absolute zero. Infra-red radiation form just part of the overall electromagnetic spectrum. Radiation is energy emitted by the electrons vibrating in the molecules at the surface of a body. The amount of energy that can be transferred depends on the absolute temperature of the body and the radiant properties of the surface.

A body that has a surface that will absorb all the radiant energy it receives is an ideal radiator, termed a "black body". Such a body will not only absorb radiation at a maximum level but will also emit radiation at a maximum level. However, in practice, bodies do not have the surface characteristics of a black body and will always absorb, or emit, radiant energy at a lower level than a black body.

It is possible to define how much of the radiant energy will be absorbed, or emitted, by a particular surface by the use of a correction factor, known as the "emissivity" and given the symbol ε. The emmisivity of a surface is the measure of the actual amount of radiant energy that can be absorbed, compared to a black body. Similarly, the emissivity defines the radiant energy emitted from a surface compared to a black body. A black body would, therefore, by definition, have an emissivity ε of 1. It should be noted that the value of emissivity is influenced more by the nature of texture of clothes, than its color. The practice of wearing white clothes in preference to dark clothes in order to keep cool on a hot summer's day is not necessarily valid. The amount of radiant energy absorbed is more a function of the texture of the clothes rather than the color.

Since World War II, there have been major developments in the use of microwaves for heating applications. After this time it was realized that microwaves had the potential to provide rapid, energy-efficient heating of materials. These main applications of microwave heating today include food processing, wood drying, plastic and rubber treating as well as curing and preheating of ceramics. Broadly speaking, microwave radiation is the term associated with any electromagnetic radiation in the microwave frequency range of 300 MHz-300 Ghz. Domestic and industrial microwave ovens generally operate at a frequency of 2.45 Ghz corresponding to a wavelength of 12.2 cm. However, not all materials can be heated rapidly by microwaves. Materials may be classified into three groups, *i.e.* conductors' insulators and absorbers. Materials that absorb microwave radiation are called dielectrics, thus, microwave heating is also referred to as dielectric heating. Dielectrics have two important properties:

- They have very few charge carriers. When an external electric field is applied there is very little change carried through the material matrix.
- The molecules or atoms comprising the dielectric exhibit a dipole movement distance. An example of this is the stereochemistry of covalent bonds in a water molecule, giving the water molecule a dipole movement. Water is the typical case of non-symmetric molecule. Dipoles may be a natural feature of the dielectric or they may be induced. Distortion of the electron cloud around non-polar molecules or atoms through the presence of an external electric field can induce a temporary dipole movement. This movement generates

friction inside the dielectric and the energy is dissipated subsequently as heat.

The interaction of dielectric materials with electromagnetic radiation in the microwave range results in energy absorbance. The ability of a material to absorb energy while in a microwave cavity is related to the loss tangent of the material.

This depends on the relaxation times of the molecules in the material, which, in turn, depends on the nature of the functional groups and the volume of the molecule. Generally, the dielectric properties of a material are related to temperature, moisture content, density and material geometry.

An important characteristic of microwave heating is the phenomenon of "hot spot" formation, whereby regions of very high temperature form due to non-uniform heating. This thermal instability arises because of the non-linear dependence of the electromagnetic and thermal properties of material on temperature. The formation of standing waves within the microwave cavity results in some regions being exposed to higher energy than others.

Microwave energy is extremely efficient in the selective heating of materials as no energy is wasted in "bulk heating" the sample. This is a clear advantage that microwave heating has over conventional methods. Microwave heating processes are currently undergoing investigation for application in a number of fields where the advantages of microwave energy may lead to significant savings in energy consumption, process time and environmental remediation.

Compared with conventional heating techniques, microwave heating has the following additional advantages:

- higher heating rates;
- no direct contact between the heating source and the heated material;
- selective heating may be achieved;
- greater control of the heating or drying process.

1.6. COMBINED HEAT TRANSFER COEFFICIENT

For most practical situations, heat transfer relies on two, or even all three modes occurring together. For such situations, it is inconvenient to analyse each mode separately. Therefore, it is useful to derive an overall heat transfer coefficient that will combine the effect of each mode within a general situation. The heat transfer in moist fabrics takes place through three modes, conduction, radiation, and the process of distillation. With a dry fabric, only conduction and radiation are present.

1.7. POROSITY AND PORE SIZE DISTRIBUTION IN FABRIC

The amount of porosity, *i.e.*, the volume fraction of voids within the fabric, determines the capacity of a fabric to hold water; the greater the porosity, the more water the fabric can hold. Porosity is obtained by dividing the total volume of water extruded from fabric sample by the volume of the sample:

Porosity = volume of water/volume of fabric
= (volume of water per gram sample)(density of sample)

It should be noted that most of water is stored between the yarns rather than within them. In the other words, all the water can be accommodated by the pores within the yarns, and it seems likely that the water is chiefly located there. It should be noted that pores of different sizes are distributed within a fabric (Figure 1.1). By a porous medium we mean a material contained a solid matrix with an interconnected void. The interconnectedness of the pores allows the flow of fluid through the fabric. In the simple situation ("single phase flow") the pores is saturated by a single fluid. In "two-phase flow" a liquid and a gas share the pore space. As it is shown clearly in Figure 1.1, in fabrics the distribution of pores with respect to shape and size is irregular. On the pore scale (the microscopic scale) the flow quantities (velocity, pressure, etc.) will clearly be irregular.

The usual way of driving the laws governing the macroscopic variables are to begin with standard equations obeyed by the fluid and to obtain the macroscopic equations by averaging over volumes or areas contained many pores.

In defining porosity we may assume that all the pore space is connected. If in fact we have to deal with a fabric in which some of the pore space is disconnected from the reminder, then we have to introduce an "effective porosity", defined as the ratio of the connected pore to total volume.

A further complication arises in forced convection in fabric which is a porous medium. There may be significant thermal dispersion, i.e., heat transfer due to hydrodynamic mixing of the fluid at the pore scale. In addition to the molecular diffusion of heat, there is mixing due to the nature of the fabric.

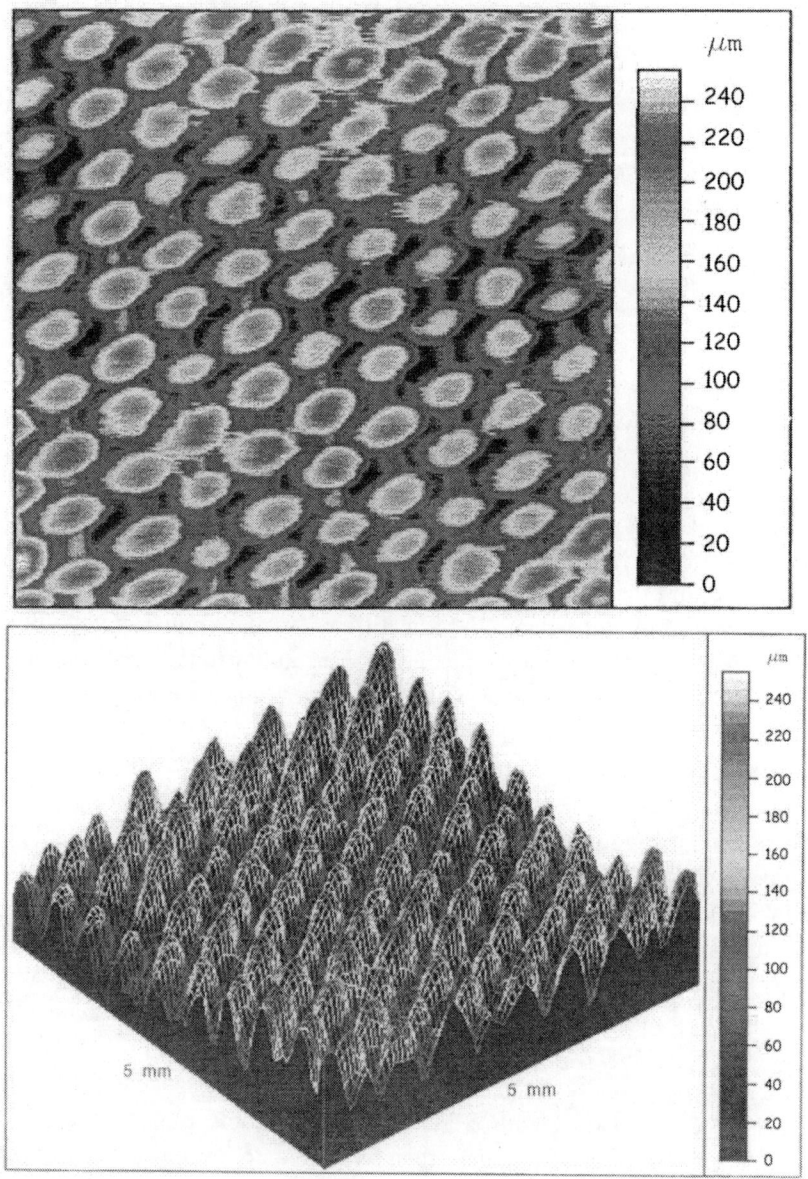

Figure 1.1. Pore size distribution within a fabric..

1.8. THERMAL EQUILIBRIUM AND COMFORT

Some of the issues of clothing comfort that are most readily involve the mechanisms by which clothing materials influence heat and moisture transfer from skin to the environment. Heat flow by conduction, convection, and radiation and moisture transfer by vapor diffusion are the most important mechanisms in very cool or warm environments from the skin.

It has been recognized that the moisture-transport process in clothing under a humidity transient is one of the most important factors influencing the dynamic comfort of a wearer in practical wear situations. However, the moisture transport process is hardly a single process since it is coupled with the heat-transfer process under dynamic conditions. Some materials will posses properties promoting rapid capillary and diffusion movement of moisture to the surface and the controlling factor will be the rate at which surface evaporation can be secured. In the initial stages of drying materials of high moisture content, also it is important to obtain the highest possible rate of surface evaporation. This surface evaporation is essentially the diffusion of vapor from the surface of the fabric to the surrounding atmosphere through a relatively stationary film of air in contact with its surface. This air film, in addition to presenting a resistance to the vapor flow, is itself a heat insulates. The thickness of this film rapidly decreases with increase in the velocity of the air in contact with it whilst never actually disappearing. The inner film of air in contact with the wet fabric remains saturated with vapor so long as the fabric surface has free moisture present. This result in a vapor pressure gradient through the film from the wetted solid surface to the outer air and, with large air movements, the rate of moisture diffusion through the air film will be considerable. The rate of diffusion, and hence evaporation of the moisture will be directly proportional to the exposed area of the fabric, inversely proportional to the film thickness and directly proportional to the inner film surface and the partial pressure of the water vapor in the surrounding air. It is of importance to note at this point that, since the layer of air film in contact with the wetted fabric undergoing drying remains saturated at the temperature of the area of contact, the temperature of the fabric surface whilst still possessing free moisture will lie very close to wet-bulb temperature of the air.

1.9. MOISTURE IN FIBERS

The amount of moisture that a fiber can take up varies markedly, as Table 1.1 shows. At low relative humidifies, below 0.35, water is adsorbed monomolecularly by many natural fibers. From thermodynamic reasoning, we expect the movement of water through a single fiber to occur at a rate that depends on the chemical potential gradient. Meanwhile, moisture has a profound effect on the physical properties of many fibers. Hygroscopic fibers will swell as moisture is absorbed and shrink as it is driven off. Very wet fabrics lose the moisture trapped between the threads first, and only when the threads themselves dry out will shrinkage begin. The change in volume on shrinkage is normally assumed to be linear with moisture content. With hydrophilic materials moisture is found to reduce stiffness and increase creep, probably as a result of plasticization. Variations in moisture content can enhance creep. To describe movement of moisture at equilibrium relative humidity below unity, the idea of absorptive diffusion can be applied. Only those molecules with kinetic energies greater than the activation energy of the moisture-fiber bonds can migrate from one site to another. The driving force for absorptive diffusion is considered to be the spreading pressure, which acts over molecular surfaces in two-dimensional geometry and is similar to the vapor pressure, which acts over three dimensional spaces [1].

Table 1.1. Smoothed values of dry-basis moisture content (kg/kg) for the adsorption of water vapor at 30°C onto textile fibers [1]

Fiber	Mc=0.2	Mc=0.5	Mc=1.0
Cotton	0.0305	0.0565	0.23
Cotton, mercerized	0.042	0.0775	0.335
Nylon 6.6, drawn	0.0127	0.0287	0.05
Orlon (50°C)	0.0031	0.0088	0.05
Cupro	0.0515	0.0935	0.36
Polyester	0.0014	0.0037	0.03
Viscose	0.034	0.062	0.25
Wool	0.062	0.09	0.38

REFERENCE

[1] R.B. Keey, The Drying of Textiles, *Rev. Prog. Coloration* 23, 57-72 (1993).

Chapter 2

CONVECTIVE HEAT FLOW

The objective of any drying process is to produce a dried product of desired quality at minimum cost and maximum throughput possible. A very common method of removing water from textiles is convective drying. Hot air is used as the heat transfer medium and is exhausted to remove vaporized water. Considerable thermal energy is required to heat make-up air as the hot air is exhausted. The effect of humidity on the drying rate of textiles depends on both gas stream temperature and flow rate, and are much larger in the warm-up and constant rate periods than in the falling rate period.

2.1. INTRODUCTION

When faced with a drying problem on an industrial scale, many factors have to be taken into account in selecting the most suitable type of dryer to install and the problem requires to be analyzed from several standpoints. Even an initial analysis of the possibilities must be backed up by a pilot-scale test unless previous experience has indicated the type most likely to be suitable. The accent today, due to high labor costs, is on continuously operating unit equipment, to what extent possible automatically controlled. In any event, the selection of a suitable dryer should be made in two stages, a preliminary selection based on the general nature of the problem and the textile material to be handled, followed by a final selection based on pilot-scale tests or previous experience combined with economic considerations.

Textile manufacturing involves a crucial energy-intensive drying stage at the end of the process to remove moisture left from dye setting. Determining drying characteristics for textiles, such as temperature levels, transition times, total drying times, and evaporation rates, is vitally important so as to optimize the drying stage. Meanwhile, a textile material undergoes some physical and chemical changes that can affect the final textile quality.

2.2. Effect of Humidity on the Drying Rate

A typical drying curve showing moisture content versus time is illustrated in Figure 2.1. It should be noted that the slope of this curve is the drying rate, at which moisture is being removed. The curve begins with a warm-up period, where the material is heated and the drying rate is usually low. As the material heats up, the rate of drying increases to a peak rate that is maintained for a period of time known as the constant rate period. Eventually, the moisture content of the material drops to a level, known as the critical moisture content, where the high rate of evaporation cannot be maintained. This is the beginning of the falling rate period. During falling rate period, the moisture flow to the surface is insufficient to maintain saturation at the surface. This period can be divided into the first and second falling rate periods. The first falling rate period is a transition between the constant rate period and the second falling rate period. In the constant rate period, external variables such as gas stream humidity, temperature, and flow rate dominate. In the second falling rate period, internal factors such as moisture and energy transport in the textile material dominate.

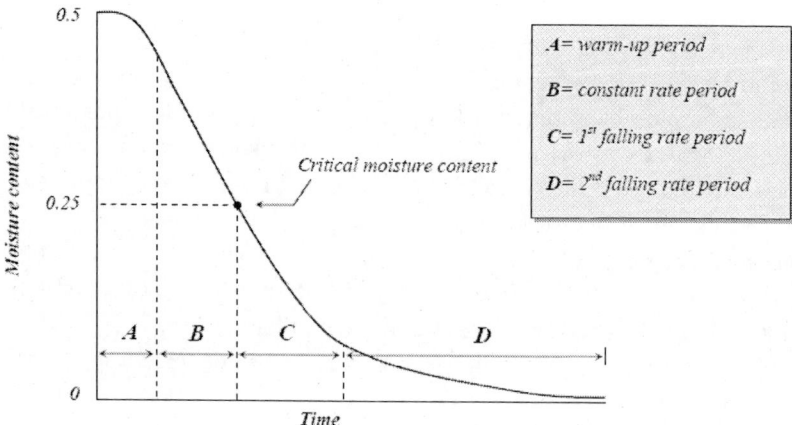

Figure 2.1. A typical moisture content profile for textile material.

Although much of the water is removed in the constant rate period of drying, the time required to reduce the moisture in the product to the desired value can depend on the falling rate period. If the target moisture content is significantly lower than the critical moisture content, the drying rates in the falling rate period become important.

The drying process can also be represented by a plot of drying rate versus moisture content, as illustrated if Figure 2.2. In this plot, time proceeds from right to left. The warm-up period is on the far right, and the constant rate period corresponds to the plateau region. The falling rate period is the section between the plateau region and the origin.

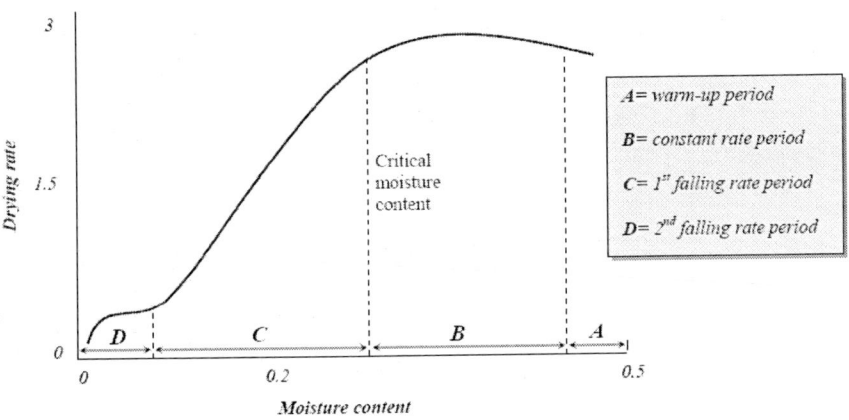

Figure 2.2. A typical drying rate profile for textile material.

2.2.1. Constant Rate Period and Falling Rate Period

During the constant rate period, evaporation is taking place from the fabric surface. The rate of drying is essentially that of the evaporation of the liquid component under the conditions of temperature and air flow during the process. High air velocities will reduce the thickness of the stationary gas film on the surface of textile material and hence increase the heat and mass transfer coefficients. In commercial forced convection dryers the effects of heat transfer by conduction and radiation may be appreciable, due to the fact that the material surface temperatures are higher than the wet-bulb temperature of the drying air.

As has been stated earlier, the period of constant rate of evaporation from wet textile material is followed by a period during which the rate of drying progressively decreases, the transition from one period to the other taking place at the point of critical moisture content of the textile material. During the constant rate drying period, surface of the exposed textile material is completely wetted, at the change to falling rate period some of the fabric surfaces will be still wet and some dry depending largely on the physical form of the textile material being dried. The rate of evaporation of the less moist surfaces will be lower than that of the completely wetted portions, the net result being a falling off in the rate of drying as drying proceeds when compared with the rate during the constant rate period. The rate of drying in this part of the drying curve will still be affected by factors which influence the constant rate drying period as discussed earlier. Once all the exposed surfaces of the textile material cease to be wetted, however, the rate of drying will be a function of the rate at which moisture or moisture vapor can move physically by diffusion and capillarity from within the fabric to its surface.

2.3. CONVECTIVE HEAT TRANSFER RATE

Many investigators have attempted to explain the effects of humidity on drying rates and the existence of inversion temperatures. The explanations are usually based on changes that occur in convective heat transfer, radiation heat transfer, and mass transfer as the humidity and temperature of the gas stream change.

At a given gas stream temperature, convective heat transfer rates can change as the humidity in the gas stream is varied, because product temperature and fluid properties vary with humidity. These effects can be explained using the following relationship for the convective heat transfer rate:

$$q/A = h(T_\infty - T_s) = h\Delta T \qquad (2.1)$$

Here q/A is the convective heat transfer rate per unit surface area A, h is the heat transfer coefficient,

T_∞ Is the free stream temperature of driving medium, T_s is the surface temperature of textile material being dried, and $\Delta T = T_\infty - T_s$ is the temperature difference between the drying medium and the textile material being dried

Since product temperature is dependent on humidity, clearly ΔT is also dependent. Further, the heat transfer coefficient h is a function of both product temperature and fluid properties. Thus the convective heat transfer rate changes with humidity, as does the drying rate of a textile material.

2.4. EQUILIBRIUM MOISTURE CONTENT

In considering a drying problem, it is important to establish at the earliest stage, the final or residual moisture content of the textile material which can be accepted. This is important in many hygroscopic materials and if dried below a certain moisture content they will absorb or "regain" moisture from the surrounding atmosphere depending upon its moisture and humidity. The material will establish a condition in equilibrium with this atmosphere and the moisture content of the material under this condition is termed the equilibrium moisture content. Equilibrium moisture content is not greatly affected at the lower end of the atmospheric scale but as this temperature increases the equilibrium moisture content figure decreases, which explains why materials can in fact be dried in the presence of superheated moisture vapor. Meanwhile, drying medium temperatures and humidity assume considerable importance in the operation of direct dryers.

In the more common drying operations met with in practice, the equilibrium moisture content of a material is important as drying may be carried out unnecessarily far, resulting in a reduction in the capacity of a given drying installation and an unjustifiably high cost of drying. Thus, if the equilibrium moisture content of wool hanks in contact with normal ambient air is of the order of 13 to 14% on the wet-weight basis there would be no point in drying to much lower moisture content. Table 2.1 gives examples of the equilibrium moisture content in contact with air at different percentage relative humidity at 16 °C.

It should be noted that two processes occur simultaneously during the thermal process of drying a wet textile material, namely, heat transfer in order to raise temperature of the wet textile material and to evaporate its moisture content together with mass transfer of moisture to the surface of the textile material and its evaporation from the surface to the surrounding atmosphere which, in convection dryers, is the drying medium. The quantity of air required to remove the moisture as liberated, as distinct from the quantity of air which will release the required amount of heat through a drop in its temperature in the course of drying, however, has to be determined from the known capacity of air to pick up moisture at a given temperature in relation to its initial content of moisture. For most practical purposes, moisture is in the form of water vapor but the same principles apply, with different values and humidity charts, for other volatile components.

Table 2.1. Equilibrium moisture content (percentage on dry-weight biases)

Material	20% r.h.a	30%r.h.a	40%r.h.a	50%r.h.a	60%r.h.a
Wool (worsted)	9.5	13.0	16.5	18.5	21.5
Cotton cloth	3.5	4.5	6.0	7.0	7.5
Egyptian cotton	3.5	4.5	5.5	6.0	7.25
Linen	2.75	3.5	4.5	5.1	6.0

r.h.a = relative humidity of air.

2.5. INVERSION TEMPERATURE

Drying air will always have an advantage over drying in steam because ΔT in Equation 2.1 is larger for drying in air; this is a consequence of T_s being very nearly the wet bulb temperature. The wet bulb temperature is lowest for dry air, increases with increasing humidity, and reaches the saturation temperature of water for a pure steam environment. Thus ΔT_{AIR} will be larger than ΔT_{STEAM}, but $\Delta T_{AIR} / \Delta T_{STEAM}$ decreases with increasing T_∞. Further, the heat transfer coefficient increases with humidity. Apparently, the net effect of the changes in h and ΔT is the convective heat transfer rate increases faster for steam than for air with increasing temperature. Some authors indicated that this is the reason inversion temperature exist. If the dominant heat transfer mechanism is convection, this explanation is plausible.

The variation of gas stream properties with humidity has been reported to explain the existence of inversion temperatures by several investigators [1, 2, 3, and 4].

For example, Sheikholeslami and Watkinson [4] explain the existence of inversion temperatures based on the total transferable heat in the drying gas. This depends on the specific heat of the gas, the mass flow rate of the gas, and

ΔT between the gas and the drying textile material. Further, Chung and Chow [1] use a dimensionless correlation for the convective heat transfer coefficient in parallel flow to explain the existence of inversion temperatures. Large Reynolds numbers and Prandl numbers favor higher drying rates in steam, while the high thermal conductivity of air tends to offset these effects. This line of reasoning may be flawed according to Sheikholeslami and Watkinson [4], who claim that convective heat transfer is altered in the presence of liquid evaporation. Thus, the correlations developed for heat transfer in the absence of drying may not be applicable to heat transfer with liquid evaporation.

Workers have proposed that in version temperatures are due to the higher radiation heat transfer in steam [5].

Differences in radiation heat transfer are grounded in the Stefan-Boltzmann law. Radiation heat transfer increases with the forth power of temperature, while convective heat transfer varies linearly with temperature. Thus, the relative importance of radiation heat transfer increases with temperature. Since the emissivity of steam is higher than that of air, radiation heat transfer from steam is greater at high temperatures.

2.6. MASS TRANSFER

In a specialized sense, the term "mass transfer" is the transport of a substance that is involved as a component (constituent, species) in a fluid mixture. An example is the transport of salt in saltine water. Moreover, convective mass transfer is analogous to convective heat transfer.

Consider a batch of fluid of volume V and mass m. Let the subscript i refer to the ith component of the mixture. The total mass is equal to the sum of the individual masses m_i so $m = \sum m_i$. Hence if the concentration of component i is defined as $C_i = m_i/V$ then the aggregate density ρ of the mixture must be the sum of all the individual concentrations, $\rho = \sum C_i$. Clearly the unit of concentration is kgm^{-3}. Instead of C_i the alternative notation ρ_i is appropriate if we think of each component spread out over the total volume V.

When chemical reactions are of interest it is convenient to work in terms of an alternative description, one involving the concept of mole. By definition, a mole is the amount of substance that contains as many molecules as there are in 12 grams of carbon 12. That number of entities is 6.022×10^{23} (Avogadro's constant). The molar mass of a substance is the mass of one mole of that substance. Hence, if there are n moles in a mixture of molar mass M and mass m, then $n = m/M$. Similarly the number of moles n_i in a mixture is the mass of that component divided by its molar mass M_i, $n_i = m_i / M_i$. The mass fraction of component i is $\phi_i = m_i / m$ so clearly $\sum \phi_i = 1$. Similarly the mole fraction of component i is $x_i = n_i / n$ and $\sum x_i = 1$. To summarize, we have three alternative ways to deal with composition-a dimensional concept (concentration) and two dimensionless ratios (mass fraction and mole fraction). These quantities are related by $C_i = \rho \phi_i = \rho(M_i / M)x_i$, where the equivalent molar mass (M) of the mixture is given by $M = \sum M_i x_i$. If, for example, the mixture can be modeled as an ideal gas, then its equation of state is $PV = mR_m T$ or $PV = nRT$, where the gas constant of the mixture (R_m) and the universal gas constant (R) are related by $R_m = n/m$, $R = R_m / M$. The partial pressure P_i of component i is the pressure we would measure if component i alone were to fill the mixture volume V at the same temperature T as the mixture. Thus $P_i V = m_i R_m T$ or $P_i V = n_i RT$. Summing these equations over i, we obtain Dalton's law, $P = \sum P_i$, which states that the pressure of a mixture of gases at a specified volume and temperature is equal to the sum of the partial pressures of the components. Note that $P_i / P = x_i$, and so using Equation $C_i = \rho \phi_i = \rho(M_i / M)x_i$ and $M = \sum M_i x_i$ we can relate C_i to P_i.

Mass transfer, as used here, is the transfer of moisture from the wet material to the gas stream. One explanation of the existence of inversion temperatures is based on the theory that the driving potential for mass transfer into an environment of steam is different from the driving potential into air [6, 7].

The driving potential for mass transfer in convective drying in air is commonly considered to be differences of vapor concentration across a boundary layer. An alternative view is that mass transfer in steam occurs by bulk flow due to a pressure difference. The vapor pressure at the surface where evaporation occurs is thought to be slightly higher than the free stream pressure, and causes bulk flow of vapor into the gas stream. Surface temperatures slightly higher than the saturation temperature have been measured [8].

Only a slight elevation in surface temperature is required to produce a pressure difference capable of generating a large bulk flow of vapor into the gas stream.

2.7. Dry Air and Superheated Steam

A number of investigators have focused on the two humidity extremes, dry air and superheated steam. Comparing drying in superheated steam with drying in air, six observations, excluding economic considerations, warrant attention. At first, the drying rate during the constant rate period [9, 10].

At saturation temperature, the drying rate in superheated steam is zero, but as the temperature increases, drying rates in superheated steam increases faster than those in air. At the inversion temperature, drying rates in superheated steam increase faster than those in air. Second, the critical moisture content decreases with increasing humidity of the drying gas. Third, in the falling rate period, drying rates in superheated steam roughly equal those in air, even at temperatures below the inversion temperature. Fourth, the equilibrium moisture content of a material dried in superheated steam at high temperatures (usually as high as the inversion temperature) is often lower than the equilibrium moisture content of a textile material dried in air at the same temperature [11, 12].

As it was mentioned earlier, equilibrium moisture content is the moisture content of a material when it comes to equilibrium with its environmental conditions, primarily the conditions of humidity and temperature. Fifth, more even moisture distribution in the textile material being dried occurs in superheated steam. When drying in air, heterogeneous surface wetting typically occurs below the critical moisture content. Sixth, material porosity and pliability are sustained better when dried in superheated steam than when dried in air.

2.8. Heat Setting Process

The main aim of the heat setting process is to ensure that fabrics do not alter their dimensions during use. This is particularly important for uses such as timing and driving belts, where stretching of the belt could cause serious problems. It is important to examine the causes of this loss in stability so that a full understanding can be obtained of the effects that heat and mechanical forces have on the stability of fabrics. All fabrics have constraints place on them by their construction and method of manufacture, but it is the heat-setting mechanism that occurs within the fiber that will ultimately influence fabric dimensions.

The temperature and time of setting must be carefully monitored and controlled to ensure the consistent fabric properties are achieved. During heat setting, the segmental motion of the chain molecules of the amorphous regions of the fiber are generally increased leading to structural relaxation within the fiber structure. During cooling, the temperature is decreased below the fiber glass transition temperature (T_g) and the new fiber structure is established.

Because the polymer chain molecules have vibrated and moved into new equilibrium positions at a high temperature in heat setting, subsequent heat treatments at lower temperatures do not cause heat-set fiber to relax and shrink, so that the fabric dimensional stability is high. Presetting of fabric prior to dyeing alters the polymer chain molecular arrangement within the fibers, and hence can alter the rate of dye uptake during dyeing. Process variations (e.g. temperature, time or tension differences) during heat setting may thus give rise to dye-ability variations that become apparent after dyeing. Fabric post-setting after coloration can lead to the diffusion of dyes such as disperse dyes to the fiber surface and sublimation, thermo-migration and blooming problems, all of which can alter the color and markedly decrease the color fastness to washing and rubbing of technical textiles containing polyester fibers.

A mixture of superheated steam and air is necessary for the rapid heat setting process. In the steam/air mixture used as the setting medium, the total heat capacity is increased considerably and thus it is possible to apply much more energy to the fabric within a given time and it is possible to heat-set the synthetic fiber fabrics in a shorter time.

An explanation for this increased energy transfer is that the specific heat of steam/air mixture is almost twice of that of air alone. Also by using steam/air mixture, heat exchange at the exterior of the fiber is increased, but the heat conduction from the surface to the interior of the fiber remains unchanged. Thus warp knitted fabrics and light weight woven like taffeta and georgette require 3-4 seconds with steam/air mixture, compared with 15 seconds if hot air alone is used. Heavier fabrics, especially woven require 9 seconds with rapid heat setting (with steam/air mixture) and 30 seconds with hot air.

Drawn polyester filaments, which have not been set have good tenacity and elasticity properties, but lack dimensional stability when subjected to the action of heat. Also, twisted or doubled filaments have a tendency to curl, which adversely affects further processing. Planar structures produced from unset continuous filaments or fibers exhibit creasing behavior when in use. It is necessary to impart dimensional stability to polyester fabrics so that the garments made from them retain their shape on being subject to washing and ironing – therefore it is necessary to set the drawn, twisted filament or fabric in this state in order to attain resistance against shrinkage, and dimensional curling and crease resistance. This can be achieved by a process called heat setting in which they are subjected to the action of heat in the presence or absence of swelling agents, with or without tension. In practice, hot water, saturated steam or dry heat is used.

The heat setting process may be explained as follows:

The linkage between the molecular chains or crystallites which were "frozen in" under tension and the mechanical stresses can be balanced by heat, thereby giving rise to greater freedom to the chains to oscillate. This provides the bonds an opportunity to snap into the sites of least energy. Further, the action of heat induces crystallization and orientation processes.

In summery, only by heat-setting the polyester fibers acquire the dimensional stability, crease-resistance and resilience desired in use. It should be noted that heat-setting is usually indispensable for ensuring satisfactory behavior of the material during other finishing processes and is one of the most important finishing processes employed with materials containing polyester fibers or their mixtures with other fibers.

2.9. CONVECTIVE HEAT AND MASS TRANSFER COEFFICIENTS

The convective heat and mass transfer coefficients at the surface of textile materials are important parameters in drying processes; they are functions of velocity and physical properties of the drying medium [13], and in general, can be expressed in the form of

$$Nu = a\,\text{Re}^b\,\text{Pr}^c \tag{2.2}$$

$$Sh = a'\,\text{Re}^{b'}\,Sc^{c'} \tag{2.3}$$

Where Nu is the Nusselt number, Re, is the Reynolds number, Pr, is the Prandtl number, S is pore saturation and $a, b, c,$ are constants.

It should be noted that for a fully wetted surface, the areas for heat and mass transfer are virtually the same, so that the surface temperature is close to the wet bulb temperature; for partly wetted surface, the effective area for mass transfer decreases with the surface moisture content.

Suzuki and Maeda [14] proposed a model for convective mass transfer coefficient which assumed that evaporation take places from discontinuous wet surface consisting of dry and wet patches. The ratio of a wet area to the total surface area decreases with decreasing moisture content. However, it is not clear how the fraction of wet area varies with the surface moisture content. Moreover for hygroscopic textile materials, when the fraction of wet area at the surface approaches zero or the surface moisture content is equal to the maximum sorptive value, the evaporation rate at the fabric surface may not be equal to zero.

2.10. CONVECTIVE DRYING OF TEXTILE MATERIAL: SIMPLE CASE

In this section we focus on the equations which consider all the major internal moisture transfer mechanisms and the properties of the textile material to be dried, including whether it is hygroscopic or non-hygroscopic. The convective heat and mass transfer coefficients are assumed to vary with the surface moisture content.

As it was mentioned earlier, drying of textile materials involves simultaneous heat and mass transfer in a multiphase system. The drying textile materials may be classified into hygroscopic and non-hygroscopic. For non-hygroscopic textile materials, pores of different sizes form a complex network of capillary paths. The water inside the pores that contribute to flow is called free water. However, the water inside very fine capillaries is difficult to replace by air. This portion of water is known as the irreducible water content. Here, it is defined as bound water. The voids in textile materials are interconnected and filled with air and certain amount of free water.

When a textile material is exposed to convective surface condition, three main mechanisms of internal moisture transfer is assumed to prevail;
- capillary flow of free water,
- movement of bound water and
- vapor transfer.

If the initial moisture content of the textile material is high enough, the surface is covered with a continuous layer of free water and evaporation takes place mainly at the surface. Internal moisture transfer is mainly attributable to capillary flow of free water through the pores. Therefore, the drying rate is determined by external conditions only, i.e. the temperature, humidity and flow rate of the convective medium, and a constant drying rate period will be observed. As drying proceeds, the fraction of wet area decreases with decreasing surface moisture content, so that the mass transfer coefficient decreases. In order to predict how the wet area fraction varies with surface moisture content, it is necessary to introduce percolation theory [15].

According to percolation theory, when water passes through randomly distributed paths in a medium, there exists a percolation threshold, which usually corresponds to critical free water movement content. When the free water content is greater than the critical, the water phase is continuous. For a two-dimensional porous medium, this critical value is about 50% of the saturated free water content and for a three-dimensional porous medium it is around 30%. Regardless of the rate of internal moisture transfer, so long as the free water content at the surface is less than the critical, the surface will form discontinuous wet patches. Thus, the mass transfer coefficient decreases with the surface free water content and the first falling rate period's starts. In the first falling rate period, a new energy balance will be reached at the surface, the 'dry' pitches still contain bound water, and the vapor pressure at the surface is determined by the Clausius-Clapeyron equation. When the surface moisture content reaches its maximum value, no free water exists. The surface temperature will rise rapidly, signaling the start of the second falling rate period, during which receding evaporation front often appears, dividing the system into two regions, the wet region and the sorption regions. Inside the evaporation front, the material is wet, i.e. the voids contain free water and the main mechanism of moisture transfer is capillary flow. Outside the front, no free water exists. All water is in the bound water state and the main mechanisms of moisture transfer are movement of bound water and vapor transfer. Evaporation takes place at the front as well as in the whole sorption region, while vapor flows through the sorption region to the surface.

Therefore, based on the definitions of the constant rate, first falling rate and second falling rate periods, the characteristics of most drying processes can be described mathematically [16].

2.10.1. Capillary Flow of Free Water

In textile material, pores provide capillary paths for free water to flow. The driving force for capillary flow is tension gradient or pressure gradient. The pertinent expression for capillary flow of free water is given by [17, 18]:

$$J_L = -\rho_W \frac{K}{\mu}(\nabla P_g - \nabla P_c - \rho_w g) \qquad (2.4)$$

Here,

J_L [$kgm^{-2}s^{-1}$] represent the free water flux, ρ_w [kgm^{-3}] is the density of water, K [m^2] refers to the permeability, μ [$kgm^{-1}s^{-1}$] is the viscosity, P_g [Nm^{-2}] is the gas pressure, P_c [Nm^{-2}] is the capillary pressure { ($P_g - P_L$)[Nm^{-2}]} and P_L [Nm^{-2}] is the liquid moisture or free water pressure.

We can also assume that:

1. the textile material is macroscopically homogeneous;
2. the flow in the capillary paths is laminar;
3. there is no significant temperature gradient;
4. the effect of the gas phase pressure and gravity force are negligible.

Thus equation (2.4) can be simplified as:

$$J_L = \rho_w \frac{K_L}{\mu}(\nabla P_g - \nabla P_c - \rho_w g) \qquad (2.5)$$

As pointed out by Miller and Miller [19], for homogeneous media and negligible gravity forces, tension is proportional to moisture content. It appears that Krischer and Kast's equation for liquid flow [20] may be valid;

$$J_L = -\rho_0 D_L \nabla U \tag{2.6}$$

where ρ_0 is the bulk density of dry material, $D_L\ [m^2 s^{-1}]$ is the capillary conductivity and $U[kgkg(solid)^{-1}]$ is the moisture content.

Since the permeability K_L depends on the pore structure of the material and the interaction between water and the textile material, it is difficult to find a theoretical form to relate K_L and the capillary conductivity D_L. However, the velocity of flow in capillaries may be assumed to follow the Hagen-Poiseuille law [21]:

$$v = \frac{r^2}{8\mu\tau}\nabla P_c \tag{2.7}$$

Where $v\,[ms^{-1}]$ is the fluid velocity, r [m] is the radius and τ is the tortuosity factor of capillary paths.

The total mass flow rate of free water can then be expressed in terms of the local velocity and corresponding capillary radius r_c, which is defined as the radius of the largest capillary in which free water exist, i.e.

$$J_L = \frac{1}{A}\int \rho_w \frac{r^2}{8\mu\tau}\nabla P_c dA_c = \int_{r_{min}}^{r_c} \rho_w \frac{r^2}{8\mu\tau}\nabla P_c \gamma(r) dr \tag{2.8}$$

where $\gamma(r)$ is the pore volume density function, defined as

$$\gamma(r) = \frac{\rho_0}{\rho_w}\frac{\Delta U}{\Delta r}\bigg|_r = \varepsilon \frac{\Delta S}{\Delta r}\bigg|_r$$

here, ε is the porosity and S is the pore saturation. The relationship between the pore saturation and the pore density function is

$$S = \frac{\int_{r_{min}}^{r_c} \gamma(r)dr}{\int_{r_{min}}^{r_{max}} \gamma(r)dr} = \frac{1}{\varepsilon} \int_{r_{min}}^{r_c} \gamma(r)dr \qquad (2.9)$$

where r_{max} is the overall largest capillary in the porous material. Comparing equations (2.5) and (2.6) with equation (2.8), K_L may be expressed as

$$K_L = -\frac{K_L}{\mu}\frac{\nabla P_c}{\nabla U} = \frac{2}{r_c^2 \gamma(r_c)}\frac{\sigma}{\mu} K_L \qquad (2.10)$$

here σ is the surface tension.

The capillary conductivity D_L may be expressed as

$$D_L = -\frac{K_L}{\mu}\frac{\rho_w \nabla P_c}{\rho_0 \nabla U} = \frac{2}{r_c^2 \gamma(r_c)}\frac{\sigma}{\mu} K_L \qquad (2.11)$$

The pore volume density function can be obtained from the relationship between capillary pressure and pore saturation of a porous medium. Solving equations (2.9) and (2.10) using the correlation of pore volume density obtained from the experimental results of Chatzis [22], a relationship between pore saturation and relative permeability can be obtained which is very close to the empirical correlation

$$K_r = \left(\frac{S - S_{ir}}{1 - S_{ir}}\right)^3 \qquad (2.12)$$

Therefore, equation (2.12) can be used to predict the relative permeability, K_r. For water, σ/μ is a linear function of temperature [23], which can be expressed as:

$$\frac{\sigma}{\mu} 1.064T - 394.3 (ms^{-1}) \qquad (2.13)$$

here T[K] is the temperature. The value of r in equation (2.11) might be constant or a function of free water content. Consequently, the following form is obtained to predict the capillary conductivity of free water for non-hygroscopic textile materials:

$$D_L = (1.604T - 394.3)\beta K_0 \left(\frac{S - S_{ir}}{1 - S_{ir}} \right)^3 \qquad (2.14)$$

here K_0 $[m^2]$ is the single phase permeability of porous material and similarly, for hygroscopic materials

$$D_L = (1.604T - 394.3)\beta K_0 \left(\frac{U - U_{ms}}{U_0 - U_{ms}} \right)^3 \qquad (2.15)$$

2.10.2. Movement of Bound Water

Movement of bound water, sometimes known as "liquid moisture transfer near dryness" or "sorption diffusion", has been studied by a number of investigators [24, 25, and 26].

It has been shown that liquid moisture transfer still exists in the sorption region and is a strong function of free water content. Whitaker and Chou [27] studied gas phase convective transport in the dry region which contains irreducible water and concluded that there could be a liquid moisture flux in the region. Movement of bound water however cannot be simply defined as a diffusion process, which often creates confusion in the analysis of liquid moisture transfer in drying processes. Moreover, the bound water conductivity measured is strongly influenced by moisture content. Therefore, movement of bound water may rather be due to flow along very fine capillaries or through cellular membranes.

In the sorption region, both movement of bound water and vapor transfer play important roles in moisture transfer. The transport equation for bound water may be expressed as

$$J_b = -\rho_w \frac{K_b}{\mu} \nabla P = -\rho_w \frac{K_b'}{\mu} \nabla P_v \qquad (2.16)$$

Since the bound water in the sorption region is in equilibrium with vapor in the gas phase, equation (2.16) may be written as

$$J_b = -\rho_w \frac{K_b'}{\mu} P_v^* \frac{\partial \psi}{\partial U} \nabla U = -\rho_0 D_b \nabla U \qquad (2.17)$$

where ψ is the relative humidity, $D_b \, [m^2 s^{-1}]$ is bound water conductivity. Thus the bound water conductivity may be written as

$$D_b = D_{bo} \left(\frac{U_b - U_{eq}}{U_{ms} - U_{eq}} \right)^3 \exp\left(\frac{E_d}{Rt} \right) \qquad (2.18)$$

here t [s] is time, E_d is defined as the activation energy of movement of bound water.

2.10.3. Vapor Flow

During a drying process, water vapor flows through the pores of the textile material by convection and diffusion. The equations of vapor flow and air flow may be written as;

$$J_v = \frac{P_v M_w}{RT} \frac{K_g}{\mu_g} \nabla P + \frac{D_v M_w}{RT} \nabla P_v \qquad (2.19)$$

$$J_a = \frac{(P - P_v) M_a}{RT} \frac{K_g}{\mu_g} \nabla P + \frac{m D_v M_a}{RT} \nabla (P - P_v) \qquad (2.20)$$

where M is the molecular weight, *m* is the ratio of air and vapor diffusion coefficients.

2.11. Macroscopic Equations Governing Heat Flow in Textile Material

The theoretical formulation of heat and mass transfer in porous media is usually obtained by a change in scale. We can pass from a microscopic view where the size of the representative volume is small with regards to pores, to a microscopic view where the size of the representative volume ω is large with regard to the pores. Moreover, the heat and mass transfer equations can be deduced from Whitaker's theory. The macroscopic equations can be obtained by averaging the classical fluid mechanics, diffusion and transfer equations over the averaging volume $\omega\,[m^3]$. The average of a function *f* is;

$$\bar{f} = \frac{1}{\omega} \int_\omega f d\omega \qquad (2.21)$$

and the intrinsic average over a phase *i* is

$$\bar{f}^i = \frac{1}{\omega_i} \int_\omega f d\omega \qquad (2.22)$$

2.11.1. Generalized Darcy's Law

Darcy's law is extended by using relative permeability. For gaseous phase, since no gravitational effect is noted;

$$\bar{v}_g = -\frac{KK_g}{\mu_g} \frac{\partial}{\partial x}(\bar{P}_g^g) \qquad (2.23)$$

where $\bar{v}_g$ is the speed of the gaseous phase, $K\,[m^2]$ the intrinsic permeability, K_g the relative permeability to the gaseous phase, $\bar{P}_g^g$ the average intrinsic pressure of the gaseous mixture, and μ_g the viscosity of the gaseous phase.

For the liquid phase

$$\overline{v}_1 = -\frac{KK_1}{\mu_1}\frac{\partial}{\partial x}(\overline{P}_g^g - \overline{P}_c^1 + \overline{\rho}_1^1 g) \tag{2.24}$$

where $\overline{v}_1$ is the speed of the liquid phase, K_1 the relative permeability to the liquid phase, P_c the capillarity pressure, μ_g µ the viscosity of the liquid, g the gravitational constant, and ρ_1 the density of liquid.

2.11.2. Mass Conservation Equations

For the liquid

$$\frac{\partial \overline{\rho}_1}{\partial t} + \frac{\partial}{\partial x}(\overline{\rho}_v^g \overline{v}_v) = \dot{m} \tag{2.25}$$

where $\dot{m}$ is the evaporated water in units of time and volume.

For the vapor

$$\frac{\partial \overline{\rho}_v}{\partial t} + \frac{\partial}{\partial x}(\overline{\rho}_v^g \overline{v}_v) = \dot{m} \tag{2.26}$$

where

$$\overline{\rho}_v^g \overline{v}_v = \overline{\rho}_v^g \overline{v}_g - \overline{\rho}_g^g D_{\mathit{eff}}\frac{\partial}{\partial x}(\overline{\rho}_v / \overline{\rho}_g) \tag{2.27}$$

P and P are the average densities of the water vapor and of the gaseous mixture, D the coefficient of the effective diffusion of vapor in the porous medium.

For the gaseous mixture

$$\frac{\partial \overline{\rho}_g}{\partial t} + \frac{\partial}{\partial x}(\overline{\rho}_g^g \overline{v}_g) = \dot{m} \tag{2.28}$$

2.11.3. Energy Conservation Equation

By assuming all the specific heats as constant and with aid of the mass conservation equations, the energy balance takes a form which is unusual but efficient in the calculations

$$\frac{\partial}{\partial t}(\overline{\rho C}_p \overline{T}) + \frac{\partial}{\partial x}(\overline{\rho}_1^l C_{p1}\overline{v}_1\overline{T} + \sum_{j=v,a}\overline{\rho}_j^g C_{pj}\overline{v}_j\overline{T})$$
$$= \frac{\partial}{\partial x}(\lambda_{eff}\frac{\partial \overline{T}}{\partial x}) - \Delta h^\circ_{vap}\dot{m} \tag{2.29}$$

here, Δh°_{vap} is a constant defined by

$$\Delta h^\circ_{vap} = \Delta h_{vap} + (C_{p1} - C_{pv})\overline{T} \tag{2.30}$$

λ_{eff} the effective thermal conductivity of the textile material, Δh_{vap} the enthalpy of vaporization, $\overline{\rho C}_p$ the constant pressure heat capacity of the textile material.

$$\overline{\rho C}_p = \overline{\rho} C_{ps} + \overline{\rho}_1 C_{p1} + \overline{\rho}_v C_{pv} + \overline{\rho}_a C_{pa} \tag{2.31}$$

2.11.4. Thermodynamic Relations

The partial pressure of the vapour is equal to its equilibrium pressure

$$\overline{P}_v^g = \overline{P}_{veq}(T,S) \tag{2.32}$$

The gaseous mixture is supported to be an ideal mixture of perfect gases

$$\overline{P}_j = \overline{\rho}_j R\overline{T}/M_j \ ; j = a, v \qquad (2.33)$$

$$\overline{P}_g = \sum_{j=a,v} \overline{P}_j \ ; \ \overline{\rho}_g = \sum_{j=a,v} \overline{\rho}_j$$
(2.34)

2.12. HEAT AND MASS TRANSFER OF TEXTILE FABRICS IN THE STENTER

By modification of the mathematical model developed by Nordon and David [28]
the transient temperature and moisture concentration distribution of a fabric in the stenter can be determined;

$$D\frac{\partial^2 C_A}{\partial x^2} = \frac{\partial C_F}{\partial t} + \varepsilon \frac{\partial C_A}{\partial t} \qquad (2.35)$$

and

$$k\frac{\partial^2 T}{\partial x^2} = \rho C_p \frac{\partial T}{\partial t} - \lambda \frac{\partial C_F}{\partial t} \qquad (2.36)$$

Here, $D[m^2/s]$ is the diffusion coefficient, $C_A[kg/m^3]$ is the moisture content of air in fabric pores, $C_F[kg/m^3]$ is the moisture content of fibers in a fabric, ε is the porosity, $k[W/mK]$ is the thermal conductivity, $\rho[kg/m^3]$ is the density, $C_p[kJ/kgK]$ is the specific heat, $\lambda[kj/kg]$ and is the latent heat of evaporation.

The boundary conditions for convective heat transfer are

$$q = h_e(T_e - T) \qquad (2.37)$$

and

$$\dot{m} = h_m(C_e - C_A) \tag{2.38}$$

here, $q[W/m^2]$ is the convective heat transfer rate, $h_e[W/m^2K]$ heat transfer coefficient, $T_e[K]$ is the external air temperature, $\dot{m}[kg/m^2s]$ is the mass transfer rate, $h_m[m/s]$ is the mass transfer coefficient, and $C_e[kg/m^3]$ is the moisture content of external air. The deriving force determining the rate mass transfer inside the fabric is the difference between the relative humidities of the air in the pores and the fibers in the fabric. The rate of moisture exchange may be considered as proportional to the relative humidity difference. Thus the rate equation for mass transfer is

$$\frac{1}{\rho(1-\varepsilon)}\frac{\partial C_F}{\partial t} = K(y_A - y_F) \tag{2.39}$$

where y_A is the relative humidity of air in pores of fabric and y_F is the relative humidity of fiber of fabric. Also, the relative humidities of air and fabric for polyester are assumed to be

$$y_A = \frac{C_A RT}{P_s} \tag{2.40}$$

and

$$y_F = \frac{C_P}{\rho(1-\varepsilon)} \tag{2.41}$$

here $R[kJ/kgK]$ is the gas constant and $P_s[kg/m^2]$ is the saturation pressure.

The rate constant K in Equation 2.39 is an unknown empirical constant. The transient fabric temperatures can be calculated assuming various values of the rate constant K. The value of the rate constant can be varied from 0.01 to

10. When the rate constant is small; the evaporation rate is so small that the moisture content decreases very slowly. Initially, the surface temperature increases rapidly, but later this rate declines (Figure 2.3).

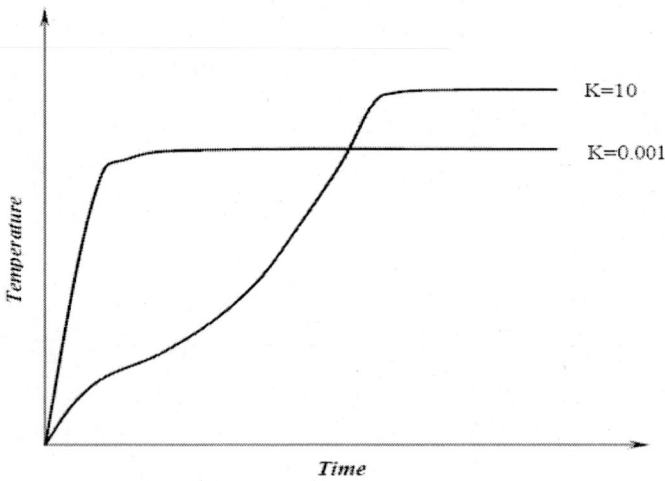

Figure 2.3. Effect of rate constant on the fabric surface temperature.

When K is greater than 1, however, the effect of the rate constant on the surface temperature distribution is not as significant. This indicates that when the rate is greater than 1, the evaporation rate is high and the diffusion mechanism inside the fabric. From Figure (2.4) we see that the surface and center temperatures increase rapidly in the initial stage up to the saturation temperature, at which point the moisture in the fabric starts to evaporate.

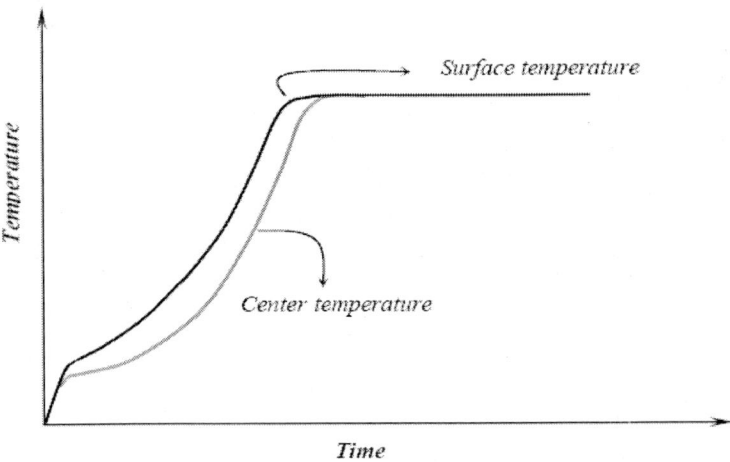

Figure 2.4. Temperature variation of surface and center fabric.

From that point, the difference between the surface temperature and the center temperature increases due to the different moisture contents of the surface and the center. In this stage the fabric stats to dry from the surface and the moisture in the interior is transferred to the fabric surface. Then the moisture content decreases during drying of the fabric. The moisture variations of the surface and the center of fabric is shown in Figure 2.5.

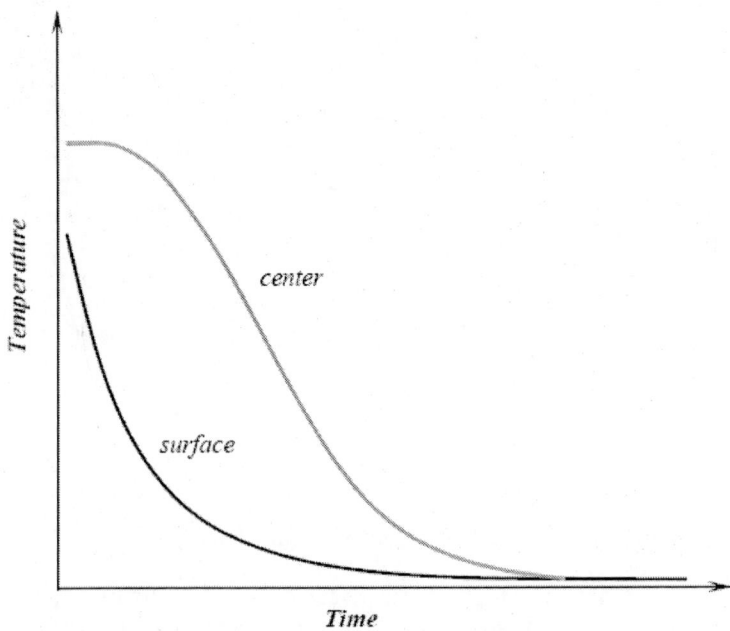

Figure 2.5. Moisture content variation of surface and center fabric.

Initially, the surface moisture content decreases rapidly, but later this rate declines because moisture is transferred to the external air from the fabric surface. The center moisture content remains constant for a short time, and then decreases rapidly, because the moisture content difference between the surface and the interior of the fabric becomes large. After drying out, both center and surface moisture contents converge to reach the external air moisture content. When the initial moisture content is high, the temperature rise is relatively small and drying takes a long time. This may be because the higher moisture content needs much more heat for evaporation from the fabric. Also, the saturation temperature for higher moisture content is lower, and thus the temperature rise in the initial stage is comparatively small (Figure 2.6).

Moreover, the effect of air moisture content can be evaluated as shown in Figure (2.7). When the moisture content is high, the initial temperature rise of the fabric also becomes high. This may be because the saturation temperature in the initial stage largely depends on the air moisture content.

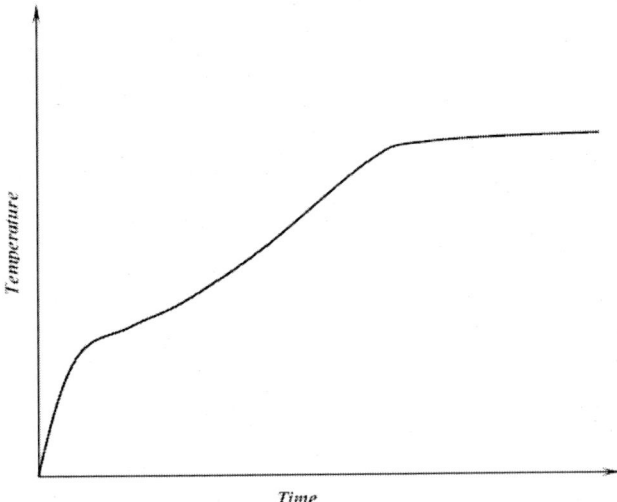

Figure 2.6. Effect of initial moisture content of fabric.

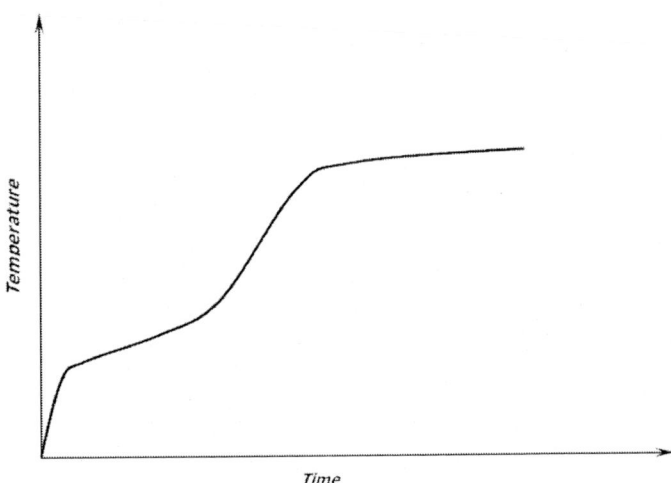

Figure 2.7. Effect of air moisture content.

After the initial temperature rise, however, the temperature increase is relatively small, and thus the time required for complete drying is comparatively long. Meanwhile, when the air flow temperature is high, the temperature rise of the fabric is great (Figure 2.8). This may be because the fabric dries more rapidly due to the large difference between its temperature and that of the air flow [29].

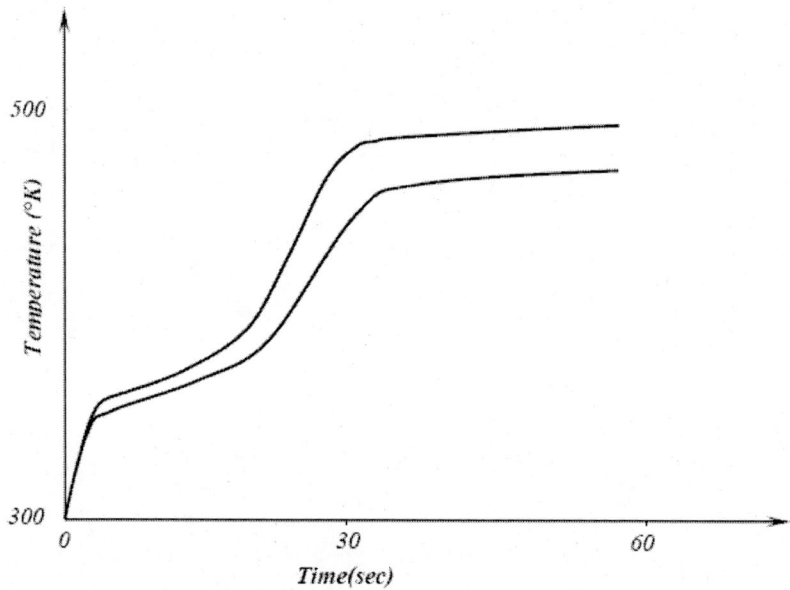

Figure 2.8. Effect of air flow temperature.

REFERENCES

[1] Chow,L. C., and Chung, J. N., Evaporation of water into laminar stream of air and superheated stem, *Int. J. Heat Mass Trans.* 26(3), 373-380 (1983).

[2] Haji, M., and Chow, L.C., Experimental measurement of water evaporation rates into air and superheated steam, *J. Heat Trans.* 110 (2), 273-242 (1988).

[3] Lane, A.M., and Stern, S., Application of superheated vapor atmospheres to drying, *Mechan. Eng.* (5), 423-426 (1956).

[4] Sheikholeslami, R., and Watkinson, A.P., rate of evaporation of water into superheated steam and humidified air, *Int. J. Heat mass Trans.* 35 (7), 1743-1751 (1992).

[5] Moyne, C., and Basilico, C., High temperature convective drying of softwood and hard wood: Drying Kinetics and Product quality interactions, *Drying '85*, 376-381 (1985).
[6] Beeby, C., and Potter, O. E., Steam Drying, Drying '85, 41-58 (1985).
[7] Shi-Ruo, C., Jin-Young, C., and Mujumdar, A.S., Preliminary study on silkworm Cocoons dring, *Drying '91*, 620-627 (1991).
[8] Wenzel, L. and White, R., Drying Granular solids in superheated steam, *Ind. Eng. Chem.* 43, 1823-1831, (1951).
[9] Nishimura, N., Nomura, T., and Ueda, S., Heat and Mass transfer with water evaporation into superheated steam stream, *Mem. Fac. Eng. Osaka city Univ.* 30, 1-10 (1989).
[10] Nomura, T., and Hyodo, T., Behavior of inversion point temperature and new applications of superheated vapor drying, *Drying '85*, 517-522 (1985).
[11] Bond, J.F., An experimental study of the falling rate period of superheated steam impingement drying of paper, *Drying Technol.* 10(4), 961-977 (1992).
[12] Trommelen, A.M., and Crosby, E.J., Evaporation and drying of drops in superheated vapours, *AICIIE J.* 16 (5), 857-867 (1970).
[13] R.B. Keey, Drying Principles and practice, 2nd edition, Pergamon Press, Oxford (1978).
[14] M.Suzuki and S. Maeda, On the mechanism of drying of granular beds, mass transfer from discountinuous source, *J. Chem. Engng Japan* 1, 26-31 (1968).
[15] R.G.Larson, L.E.Scriven and H.T. Davis, Percolation theory of two-phase flow in porous media, *Chem. Engng Sci.* 36, 57-73 (1981).
[16] P.Chen and D.C.Pei, A mathematical model of drying processes, *Int J. Heat Mass Transfer*, 32(2), 297-310, 1989.
[17] R.A. Greenkorn, Flow phenomena in porous media, *Marcel Dekker*, Newyork (1983).
[18] S. Whitaker, Flow in porous media II: the governing equations for immiscible, two-phase flow, *Transport porous media 1*, 105-125 (1986).
[19] E.E. Miller and R.D. Miller, Theory of capillary flow I, Practical implications, *Proc. Soil Sci. Soc. Am.* 19, 267-271 (1995).
[20] O. Krischer and W. Kast, Die Wissenschaftlichen Grundlagen der Trocknungstechnik, 3rd Edn, Springer, Berlin (1978).
[21] R.B. Keey, Drying principles and practice, 1st edition Pergamon press, Oxford (1972).

[22] I.Chatzis, A network approach to analyse and model capillary and transport phenomena in porous media, Ph.D. Thesis, University of waterloo, Canada (1980).
[23] P. Chen, Mathematical modeling of drying and freezing processes in the food industry, PhD. Thesis, University of Waterloo, Canada, (1987).
[24] E. Rotstein, Advances in transport phenomena and thermodynamics in the drying of cellular food systems. In Drying '86 (Edited by A.S. Mujumdar), Vol. 1, pp. 1-11. Hemisphere, Washington, DC (1986).
[25] E. Rotstein and A.R.H. Cornish, Influence of cellular membrane permeability on drying behavior, *J. Food Sci.* 43, 926-934 (1978).
[26] G. Bramhall, Sorption diffusion in wood, Wood Sci. 12, 3-13 (1979).
[27] S. Whitaker and W.T. Chou, Drying granular porous media-theory and experiment, *Drying Tech. Int. J.* 1, 3 (1983).
[28] Nordon, P., and David, H. G., Coupled diffusion of moisture and heat in Hygroscopic textile materials, *Int. J. Heat Mass Trans.* 10, 853-866 (1967).
[29] S. Park and D. Baik, Heat and mass transfer analysis of fabric in tenter frame, *Textile Res. J.* 67(5), 311-316 (1997).

Chapter 3

CONDUCTIVE HEAT FLOW

Conduction is the process of heat transfer by molecular motion, supplemented by the flow of heat through textile material from a region of high temperature. Heat transfer by conduction takes place across the interface between two bodies in contact when they are at different temperatures. A common example of heat conduction is heating textile fabric in a cylindrical dryer.

3.1. INTRODUCTION

In most textile engineering problems, our primary interest lies not in the molecular behavior of textiles, but rather in how the textile behaves as a continuous medium. In our study of heat conduction, we will therefore neglect the molecular structure of the textile and consider it to be a continuous medium-continuum, which is a valid approach to many practical problems were only macroscopic information is of interest. Such a model may be used provided that the size and the free path of molecules are small compared with other dimensions existing in the medium, so that a statistical average is meaningful. This approach, which is also known as the phenomenological approach to heat conduction, is simpler than microscopic approaches and usually gives the answers required in textile engineering. For heat conduction problems, the use of first and second laws of thermodynamics is sufficient. In addition to these general laws, it is usually necessary to bring certain particular laws into an analysis. There are three such particular laws we employ in the analysis of conduction heat transfer:

a) Fourier's law of heat conduction
b) Newton's law of cooling, and
c) Stefan-Boltzmann's law of radiation.

3.2. FIRST LAW OF THERMODYNAMICS

When a system undergoes a cyclic process, the first law of thermodynamics can be expressed as

$$\oint \delta Q = \oint \delta W \qquad (3.1)$$

where cyclic integral $\oint \delta Q$ represents the net heat transferred to the system, and the cyclic integral $\oint \delta W$ is the net work done by the system during cyclic process. Both heat and work are path functions. For a process that involves an infinitesimal change of state during a time interval dt, the first law of the thermodynamics is given by

$$dE = \delta Q - \delta W \qquad (3.2)$$

where δQ and δW are the differential amounts of heat added to the system and the work done by the system, respectively, and dE is the corresponding increase in the total energy of the system during the time interval dt. The energy E is a property of the system and, like other properties, is a point function. That is, dE, depends on the initial and final states only, and not on the path followed between the two states. The physical property E represents the total energy contained within the system and is customarily separated into three parts as bulk kinetic energy, bulk potential energy, and internal energy; that is,

$$E = KE + PE + U \qquad (3.3)$$

The internal energy U, which includes all forms of energy in a system other than bulk kinetic and potential energies, represents the energy associated with molecular and atomic structure and behaviour of the system.

Equation (3.2) can also be written as a rate equation:

$$\frac{dE}{dt} = \frac{\delta Q}{dt} - \frac{\delta W}{dt} \qquad (3.4a)$$

or

$$\frac{dE}{dt} = q - \dot{W} \qquad (3.4b)$$

where $q = \delta Q/dt$ represents the rate of heat transfer to the system and $\dot{W} = \delta W/dt$ is the rate of work done by the system.

3.3. SECOND LAW OF THERMODYNAMICS

The second law leads to the thermodynamic property of entropy. For any reversible process that a system undergoes during a time interval dt, the change in the entropy S of the system is given by

$$dS = \left(\frac{\delta Q}{T}\right)_{rev} \qquad (3.5a)$$

For an irreversible process, the change, however, is

$$dS \succ \left(\frac{\delta Q}{T}\right)_{irr} \qquad (3.5b)$$

where δQ is the small amount of heat added to the system during the time interval dt, and T is the temperature of the system at the time of heat transfer. Equations (3.5) may be taken as the mathematical statement of the second law, and they can also be written in rate form as

$$\frac{dS}{dt} \geq \frac{1}{T}\frac{\delta Q}{dt} \qquad (3.6)$$

The control-volume from of the second law can be developed also by:

$$\frac{\partial}{\partial t}\int_{cv} s\rho dv + \int_{cs} s\rho V.\hat{n}dA \geq \int_{cs}\frac{1}{T}\frac{\delta Q}{\delta t} \qquad (3.7)$$

where s is the entropy per unit mass, and the equality applies to reversible processes and the inequality to irreversible processes.

3.4. HEAT CONDUCTION AND THERMAL CONDUCTIVITY

The rate of heat conduction through a material may be proportional to the temperature difference across the material and to the area perpendicular to the heat flow and inversely proportional to the length of the path of heat flow between the two temperature levels. This dependence was established by well known French scientist J.B.J. Fourier, who used it in his remarkable work, *Theorie Analytique de la Chaleur*, published in Paris in 1822. In this book he gave a very complete exposition of the theory of heat conduction. The constant of proportionality in Fourier's law, denoted by k, is called the thermal conductivity. It is a property of the conducting material and of its state. With the notation indicated in figure 3.1, Fourier's law is

$$q = \frac{kA}{L}(t_1 - t_2) \qquad (3.8)$$

where kA/L is called the conductance of the geometry. In figure 3.1, since there is a temperature difference of $(t_1 - t_2)$ between the surfaces, heat will flow through the material. From the second law of thermodynamics, we know that the direction of this flow is from the higher temperature surface to the lower one. According to the first law of thermodynamics, under steady conditions, this flow of heat will be at a constant rate.

The thermal conductivity k, which is analogous to electrical conductivity, is a property of the thermal material. It is equivalent to the rate of heat transfer between opposite faces of a unit cube of the material which are maintained at temperatures differing by 1°. In SI unit, k is expressed as W/mK.

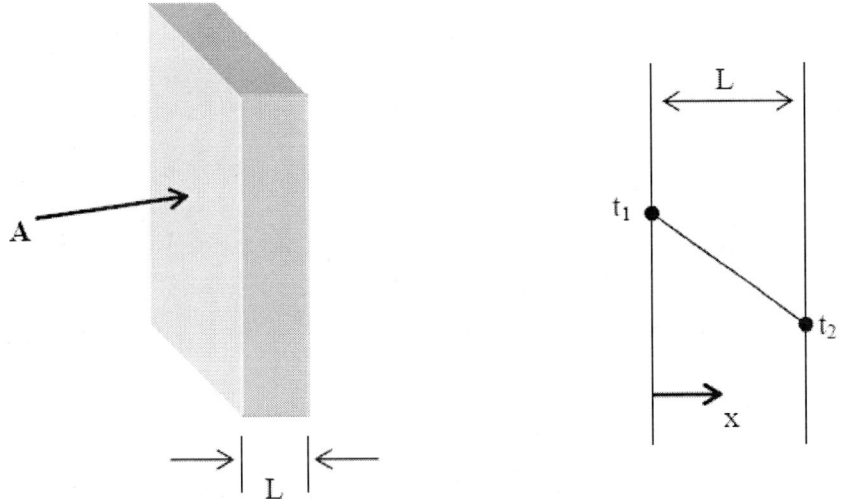

Figure 3.1. One-dimensional steady-state heat conduction.

The conduction equation 3.8 may also be written as the heat transfer rate per unit area normal to the direction of heat flow, q'', as

$$\frac{q}{A} = q'' = \frac{k}{L}(t_1 - t_2) = k\left[-\frac{(t_2 - t_1)}{L}\right] = -k\frac{dt}{dx} \qquad (3.9)$$

The quantity q'' is very useful and is hereafter called the heat flux. The quantity in the brackets is minus the temperature gradient through the material, that is, $-dt/dx$.

Moreover, thermal conductivity is a thermo-physical property. The thermal conductivity of a material depends on its chemical composition, physical structure, and state. It is also varies with the temperature and pressure to which the material is subjected. In most cases, however, thermal conductivity is much less dependent on pressure than on temperature, so that the dependence on pressure may be neglected and thermal conductivity can be tabulated as a function of temperature. In some case, thermal conductivity may also vary with direction of heat flow as in anisotropic materials. The variation of thermal conductivity with temperature may be neglected when the temperature range under consideration is not too severe. For numerous materials, especially within a small temperature range, the variation of thermal conductivity with temperature can be presented by the linear function

$$k(T) = k_0\left[1 + \gamma(T - T_0)\right] \tag{3.10}$$

where $k = k(T_0)$; T_0 is a reference temperature, and γ is a constant called the temperature coefficient of thermal conductivity.

Heat conduction in gases and vapors depends mainly on the molecular transfer of kinetic energy of the molecular movement. That is, heat conduction is transmission of kinetic energy by the more active molecules in high temperature regions to the molecules in low molecular kinetic energy regions by successive collisions. According to kinetic theory of gases, the temperature of an element of gas is proportional to the mean kinetic energy of its constituent molecules. Clearly, the faster the molecules move, the faster they will transfer energy. This implies, therefore, that thermal conductivity of a gas should be dependent on its temperature.

3.5. THERMAL CONDUCTION MECHANISMS

Theoretical predictions and measurements have been made of the value of thermal conductivity, k, for many types of substances. In gases, heat is conducted (*i.e.*, thermal energy is diffused) by random motion of molecules. Higher-velocity molecules from higher-temperature regions move about randomly, and some reach regions of lower temperature. By a similar random process, lower-velocity molecules from lower-temperature regions reach higher temperature regions. Thereby, net energy is exchanged between the two regions. The thermal conductivity depends upon the space density of molecules, upon their mean free path, and upon the magnitude of the molecular velocities. The net result of these effects, for gases having very simple molecules, is dependence of k upon **T**, where T is the absolute temperature. This results from kinetic of gases.

3.6. MASS DIFFUSION AND DIFFUSIVITY

Mass diffusion through a region occurs by the motion of quantities of mass or chemical species through the textile material. The diffusion mechanism in a plane layer of textile material is formulated in Figure 3.2.

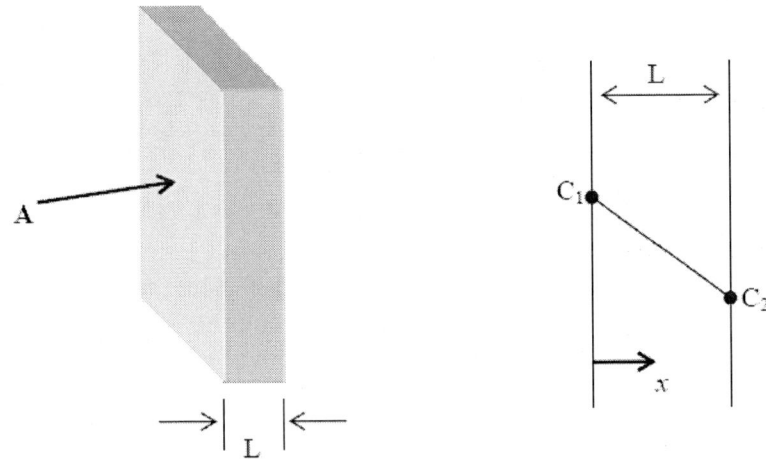

Figure 3.2. Imposed surface concentration C_1 and C_2 at $x=0$ and L, for a diffusing textile material.

A layer L thick and of area A has concentration levels C_1 and C_2 maintained at the two boundaries, at $x=0$ and $x=L$. In this example it is assumed that there is no local adsorption or release of stored or bound diffusing material C in the region $x=0$ to L. The mass diffusion coefficient D is also taken as uniform, that is, constant across the region.

The rate of mass diffusion through the layer is steady state, m, will be proportional to the concentration difference across the material, $(C_1 - C_2)$, and to the area perpendicular to the mass diffusion, A. It will be inversely proportional to the length of the path L of mass diffusion between the two imposed concentration levels C_1 and C_2. Then the mass flow rate through the region, m, of species C is given by

$$m = \frac{DA}{L}(C_1 - C_2) \qquad (3.10)$$

where DA/L is the "mass" conductance across the region. This result is completely analogous to Equation (3.8) for heat flow, which follows from the FOURIER LAW OF CONDUCTION. The mass diffusion formulation given in Equation (3.10) is a form of Fick's first law.

Equation (3.10) is also written in terms of a mass flux, that is, the diffusion rate in mass per unit cross-section area, per unit time, as m'':

$$\frac{m}{A} = m'' = \frac{D}{L}(C_1 - C_2) = D\left[-\frac{(C_2 - C_1)}{L}\right] = -D\frac{dC}{dx} \quad (3.11)$$

The quantity in brackets is minus the concentration gradient through the textile material, that is $-dC/dx$.

Several consistent systems of dimensions and units are used for the physical quantities in Equation (3.11). A common practice expresses concentration in mass per unit volume, M/L^3, m'' in mass flux per unit area and time, that is, M/L^2T. Then D has the dimensions of L^2/T. The unit may be cm^2/s or ft^2/hr, or as m^2/s in SI units. Other commonly used measures and terms of composition and concentration include number density of fraction, partial density, mass or mole fraction, and molar concentration.

The magnitude of the diffusion coefficient depends strongly on both the material through which diffusion occurs and the diffusing species. As for heat conduction, as discussed in Equation (3.11) indicates that both the gradient of concentration dC/dx and the mass flux $m''(x)$ are constant across the region. This follows from the condition of steady state when D is uniform across the region and no local adsorption of species C occurs in the region.

3.7. CONDUCTION HEAT TRANSFER IN TEXTILE FABRIC

In the cylindrical dryers, the transfer of heat to textile fabrics can be calculated by

(Heat transferred to fabric) = (Heat absorbed by fabric) − (Heat transferred from fabric to the ambient air by convection and radiation)

That means;

$$\alpha(T_\infty - T)(dA)(dt) = (m_s)(C)(dA)(dt) + \alpha_t(T - T_a)(dA)(dt)$$

or

$$\alpha(T_\infty - T)dt = (m_s)(C)(dT) + \alpha_t(T - T_a)dt \qquad (3.12)$$

Where

$\alpha[Wm^{-2}K^{-1}]$ is the overall heat transfer coefficient of cylindrical dryer-fabric, T[K] is the relative temperature of fabric, $T_\infty[K]$ is the temperature of cylindrical dryer, dA is the differential surface of fabric, t[s] is the heating period, $m_s[kg.m^{-2}]$ is the mass of fabric, $T_a[K]$ is the temperature of ambient and $C[J.kg^{-1}K^{-1}]$ is the specific heat of textile fabric.

It should be noted that thermal loss is an important parameter which may occur during high rotation of the cylinders. Meanwhile, α_t is a heat transfer coefficient which characterizes the thermal loss of fabric toward the ambient. This can be shown in the general form of

$$\alpha_t = a(T - T_a) + b \qquad (3.13)$$

Substitution of Equation (3.13) into Equation (3.12) yields

$$\alpha(T_\infty - T)dt = (m_s)(C)dT + \big(a(T - T_a) + b\big)(T - T_a)dt \qquad (3.14)$$

We assume that α is independent of temperature, then the Equation (3.14) can be written as:

$$T^2 + K_1 T + K_2 \frac{dT}{dt} = K_3 \qquad (3.15)$$

with

$$K_1 = \frac{(\alpha + b - 2aT_a)}{a}, \quad K_2 = \frac{(m_s)(C)}{a}, \text{ and}$$

$$K_3 = \frac{\alpha T_\infty + bT_a - aT_a^2}{a}$$

meanwhile, $T_1 = \theta$ could be a particular solution which corresponds to equilibrium temperature $\left(\dfrac{dT_1}{dt}\right) = 0$.

Therefore,

$$\theta^2 + K_1\theta - K_3 = 0$$

$$\theta_1 = \frac{-K_1 + \sqrt{\delta'}}{2} \tag{3.16}$$

$$\theta_2 = \frac{-K_1 - \sqrt{\delta'}}{2} \tag{3.17}$$

with

$$\delta' = K_1^2 + 4K_3 \tag{3.18}$$

and

$$T = \theta + \frac{1}{(\beta)\exp\left(K_2^{-1}(K_1 + 2\theta)t\right) - (K_1 + 2\theta)^{-1}} \tag{3.19}$$

To identify θ, we can show,

$$\lim_{t \to \infty} T = \theta$$

θ is equal to the equilibrium temperature of fabric i.e., $\theta = T_{eq}$.

and

$$\beta = (T_a - T_{eq})^{-1} + (K_1 + 2T_{eq})^{-1} \tag{3.20}$$

REFERENCES

[1] S.Kakac and Y. Yener, *Heat conduction,* Taylor & Francis Pyblications, 1993.

[2] B.Gebhart, *Heat conduction and Mass Diffusion,* Mc Graw-Hill, 1993.

Chapter 4

RADIATIVE HEAT FLOW

Radiation is a form of electromagnetic energy transmission and takes place between all matters providing that it is at a temperature above absolute zero. Infra-red radiation form just part of the overall electromagnetic spectrum. Radiation is energy emitted by the electrons vibrating in the molecules at the surface of a body. The amount of energy that can be transferred depends on the absolute temperature of the body and the radiant properties of the surface.

Electromagnetic radiation is a form of energy that propagates through a vacuum in the absence of any moving material. We observe electromagnetic radiation as light and use it as radio waves, X-rays, etc. Here, we are mostly interested in a form of electromagnetic radiation called microwaves that can be used to heat and dry textile materials.

4.1. INTRODUCTION

The word *microwave* is not new to every walk of life as there are more than 60 million microwave ovens in the households all over the world. On account of its great success in processing food, people believe that the microwave technology can also be wisely employed to process materials. Microwave characteristics that are not available in conventional processing of materials consist of: penetrating radiation, controllable electric field distribution, rapid heating, selective heating materials and self-limiting reactions. Single or in combination, these characteristics lead to benefits and opportunities that are not available in conventional processing methods.

Since World War II, there have been major developments in the use of microwaves for heating applications. After this time it was realized that microwaves had the potential to provide rapid, energy-efficient heating of materials. These main applications of microwave heating today include food processing, wood drying, plastic and rubber treating as well as curing and preheating of ceramics. Broadly speaking, microwave radiation is the term associated with any electromagnetic radiation in the microwave frequency range of 300 MHz-300 Ghz. Domestic and industrial microwave ovens generally operate at a frequency of 2.45 Ghz corresponding to a wavelength of 12.2 cm. However, not all materials can be heated rapidly by microwaves. Materials may be classified into three groups, *i.e.* conductor insulators and absorbers. Materials that absorb microwave radiation are called dielectrics, thus, microwave heating is also referred to as dielectric heating. Dielectrics have two important properties:

- They have very few charge carriers. When an external electric field is applied there is very little change carried through the material matrix.
- The molecules or atoms comprising the dielectric exhibit a dipole movement distance. An example of this is the stereochemistry of covalent bonds in a water molecule, giving the water molecule a dipole movement. Water is the typical case of non-symmetric molecule. Dipoles may be a natural feature of the dielectric or they may be induced. Distortion of the electron cloud around non-polar molecules or atoms through the presence of an external electric field can induce a temporary dipole movement. This movement generates friction inside the dielectric and the energy is dissipated subsequently as heat [1].

The interaction of dielectric materials with electromagnetic radiation in the microwave range results in energy absorbance. The ability of a material to absorb energy while in a microwave cavity is related to the loss tangent of the material.

This depends on the relaxation times of the molecules in the material, which, in turn, depends on the nature of the functional groups and the volume of the molecule. Generally, the dielectric properties of a material are related to temperature, moisture content, density and material geometry.

An important characteristic of microwave heating is the phenomenon of "hot spot" formation, whereby regions of very high temperature form due to non-uniform heating. This thermal instability arises because of the non-linear dependence of the electromagnetic and thermal properties of material on temperature. The formation of standing waves within the microwave cavity results in some regions being exposed to higher energy than others. Cavity design is an important factor in the control, or the utilization of this "hot spots" phenomenon.

Microwave energy is extremely efficient in the selective heating of materials as no energy is wasted in "bulk heating" the sample. This is a clear advantage that microwave heating has over conventional methods. Microwave heating processes are currently undergoing investigation for application in a number of fields where the advantages of microwave energy may lead to significant savings in energy consumption, process time and environmental remediation.

The benefit of microwave technology has been realized over the past decade with the growing acceptance of microwave ovens in the home. This, together with the gloomy outlook of worldwide energy crises, has paved the way for extensive research into new and innovative heating and drying processes. The use of microwave drying cannot only greatly enhance the drying rates of textile materials, but it may also enhance the final product quality.

While cost presents a major barrier to wider use of microwave in textile industry, an equally important barrier is the lack of understanding of how microwaves interact with materials during heating and drying. The design of suitable process equipment is further confounded by the constraint that geometry places on the prediction of field patterns and hence heating rates within the materials. Effects such as resonance within the material can occur as well as large variations in field patterns at the textile material surface.

The phenomenon of drying has been investigated at considerable length and treated in various texts. However in general, there is only a very small section of this literature devoted to microwave to microwave drying of textile materials.

One of the main features which distinguish microwave drying from conventional drying processes is that because liquids such as water absorb the bulk of the electromagnetic energy at microwave frequencies, the energy is transmitted directly to the wet material. The process does not rely on conduction of heat from the surface of the textile material and thus increased heat transfer occurs, speeding up the drying process. This has the advantage of eliminating case hardening of textile material which is usually associated with convective hot air drying operations. Another feature is the large increase in the dielectric loss factor with moisture content. This can be used with great effect to produce a moisture leveling phenomenon during the drying process since the electromagnetic energy will selectively or preferentially dry the wettest regions of the solid [2].

Meanwhile, infrared heating on textile lines has been in use for many years on dyeing lines to pre-dry a host of fabric finishes or topical coatings on fabrics. The renewed interest in infrared pre-drying is due in large part to the need for ever-increasing line speeds and the availability of improved infrared hardware. Infrared predrying of the dyed or finished fabric rapidly preheats and pre-dries wetted fabrics far faster than the typical convection dryer. Typically an air dryer requires 20-25% of its length just to preheat the wetted fabric to a temperature where water is freely evaporated. The infrared pre-heater/pre-dryer section takes over this function in a fraction of the length required in the convection dryer. For dyed fabrics, infrared pre-dryers are typically vertical in configuration, and are generally mounted on the line prior to the drying frame. The systems consist of arrays of electric infrared emitters positioned on both sides of the fabric. The emitters are typically controlled from the fabric temperature. The evaporative load on the pre-dryer dictates how much energy is required and how many vertical sections the pre-dryer must be. With today's more efficient and higher powered emitters most pre-dryers are one or two passes.

It should be noted that controlling shade variations and shade shifts in dyed fabrics has typically been problematic for manufacturing engineers. Dyestuffs tend to migrate to the heated side of the fabric as it passes through the oven. The migration is due partly to gravity, and partly to fluid dynamics. Dyed fabrics come onto the dryer frame at usually 50% to 80% wet pickup. Optimum product quality requires that wet pickup be reduced to the 30% to 60% range with equal water removal from both sides of the fabric. The pre-dried fabric is then presented to the horizontal drying oven with the dyes "locked in" to position. Additional quality benefits can be realized on topical finishes or coatings. Rapid heating with infrared immediately after coating applications tends to keep the coating from deeply wicking into the fabric [3].

4.2. BACKGROUND

This section reviews the basic principles of physics pertaining to microwave heating.

Energy: energy is the capacitance to do work, and work is defined as the product of force acting over a distance, that is,

$$E = W = (F)(x) \tag{4.1}$$

Where E = energy, W = the equivalent work, F = force that performs the work, x = distance a mass is moved by the force.

Atomic particles: all matter are composed of atoms. Atoms, in turn, consist of a nuclei surrounded by orbiting electrons. The nucleus consists of positively charged protons and unchanged neutrons. The surrounding electrons are negatively charged. In neutral atoms, the number of protons in the nucleus equals the number of electrons, resulting in a 0 net charge.

Electrostatic forces: if some electrons are removed from a piece of material, the protons will outnumber the electrons and the material will take on a positive charge. Similarly, if some electrons are added to a piece of material, the material will take a negative charge. If two positively charged objects are brought near to each other, they will each feel a force pushing them apart. Similarly, if two negatively charged objects are brought together, they will each experience a force pushing them apart. On the other hand, if a negatively charged object is brought near a positively charged object, each will experience a force pulling them together.

Columb's law: if two charges of magnitude q_1 and q_2 are separated by a distance r as shown if Figure 4.1, each will feel a force magnitude ;

$$F = k \frac{q_1 q_2}{r^2} \tag{4.2}$$

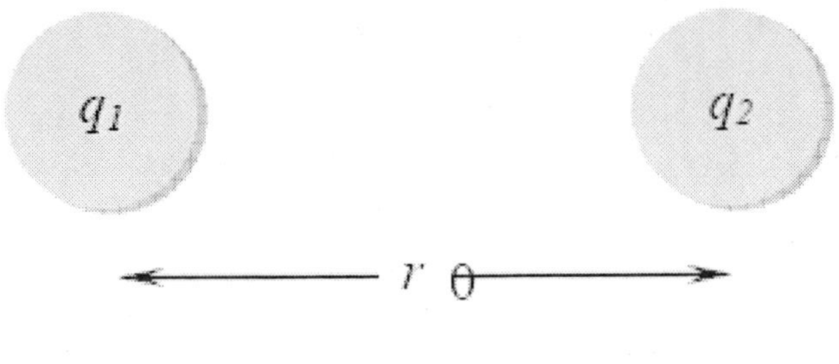

Figure 4.1. Principle of Coulomb's Law.

It is clear from Equation (4.2) that the force is proportional to the magnitude of each charge and inversely proportional to the square of the distance between them. If, for example, we double the charge on either object, the force will double. On the other hand, if we double the distance between them, the force will be reduced to 1/4 of its previous value.

Electric fields: Electrostatic force is defined as "force at a distance" (Equation 4.1). If we have a charge Q, and a test charge q is placed a distance R away from it, Q will push on q across that distance as shown in Figure 4.2. The magnitude of push will depend on the magnitudes of Q, q, and r as given in Equation (4.2).

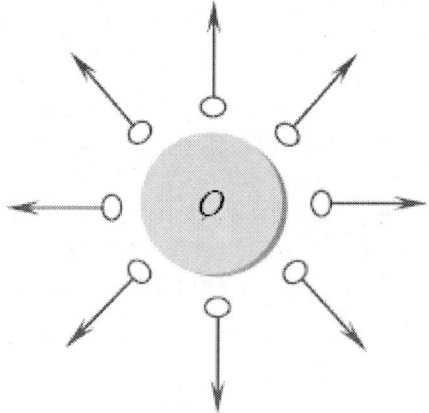

Figure 4.2. Forces around charge Q.

Another way to look at this is to say that Q creates a field in the space that surrounds it. At any point in that space, the field will have a strength E that depends on Q and r. If a test charge is placed at some point in the space, the field at that point will push on it with a force depends on the field strength E at that point and on q. To make these two explanations mathematically equivalent, we separate Equation (4.2) into two parts; thus

$$F = k\frac{Qq}{r^2} \left(k\frac{Q}{r^2} \right)(q) \qquad (4.3)$$

The second part is simply the charge of the second particle. The first part we call E, the field strength at distance r away from Q:

$$E = \left(k\frac{Q}{r^2}\right) \qquad (4.4)$$

Now the force on q can be defined in terms of the field strength times the magnitude of q:

$$F = E.q \qquad (4.5)$$

A microwave oven consists of three major parts:

- The magnetron is the device that generates the microwaves.
- Wave guides direct these waves to the oven cavity.
- The oven cavity holds the material to be heated so that microwaves can impinge on them.

Magnetron: It generates microwaves and consists of the following parts:

a) Central cathode. The cathode is a metal cylinder at the center of the magnetron that is coated with an electron-emitting material. In operation, the cathode is heated to a temperature high enough to cause electrons to boil off the coating.
b) Outer anode. There is a metal ring called an anode around the magnetron that is maintained at a large positive potential (voltage) relative to the cathode. This sets up an electrostatic field between the cathode and anode that accelerates the electrons toward the anode.

Magnetic field: a strong magnetic field is placed next to the anode and cathode in such an orientation that it produces a magnetic field at right angles to the electrostatic field. This field has the effect of bending the path of the electrons so that, instead of rushing to the anode, they begin to circle in the space between the cathode and anode in a high-energy swarm.

Resonant cavities: they have been built into the anode. Random noise in the electron swarm causes occasional electrons to strike these cavities is such that most radiation frequencies die out. Microwave frequencies, on the other hand, bounce around the cavities and tend to grow, thus getting their energy from the magnetron, passes through the wave guides, and enters the cavity.

However, not all materials can be heated rapidly by microwaves. Materials are reflected from the surface and therefore do not heat metals.

Metals in general have high conductivity and are classed as conductors. Conductors are often used as conduits (waveguide) for microwaves. Materials which are transparent to microwaves are classed as insulators. Insulators are often used in microwave ovens to support the material to be heated. Materials which are excellent absorbers of microwave energy are easily used and are classed as dielectric. Figure 4.3 shows these properties.

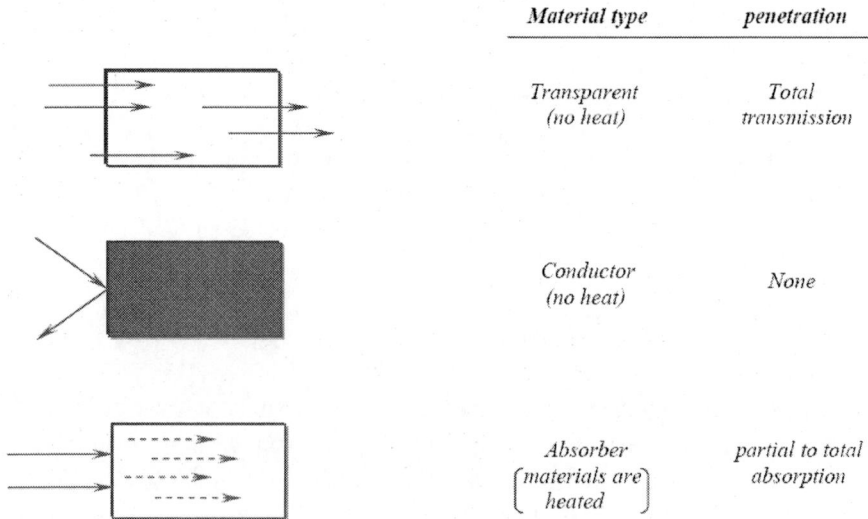

Figure 4.3. Interaction of microwave with materials.

Table 4.1 shows the electromagnetic spectrum. In this continuum, the radio-frequency range is divided into bands as depicted in Table 4.2. Radio-frequency (r.f.) energy has several possible benefits in textile processing. Substitution of conventional heating methods by radio-frequency techniques may result in quicker and more uniform heating, more compact processing machinery requiring less space, and less material in-process at a particular time. Radio-frequency energy has been used for many years to heat bulk materials such as spools of yarn. Bands 9, 10, and 11 constitute the microwave range that is limited on the frequency side by HF and on the high frequency side by the infrared. These microwaves propagate through empty space through empty space at the velocity of light. The frequency ranges from 300 MHz to 300 GHz.

Table 4.1. The electromagnetic spectrum

Region	Frequencies (Hz)	Wavelength
Audio frequencies	$30 - 30 \times 10^3$	10mm-10km
Radio frequencies	$30 \times 10^3 - 30 \times 10^{11}$	10km-1m
Infrared	$30 \times 10^{11} - 4 \times 10^{14}$	1m-730nm
Visible	$4 \times 10^{14} - 7.5 \times 10^{14}$	730nm-0.3nm
Ultraviolet	$7.5 \times 10^{14} - 1 \times 10^{18}$	400nm-0.3nm
X-rays	$> 1 \times 10^{17}$	< 3nm
Gamma rays	$> 1 \times 10^{20}$	< 3nm
Cosmic rays	$> 1 \times 10^{20}$	< 3nm

Table 4.2. Frequency bands

Band	Designation	Frequency limits
4	Very low frequency (VLF)	3-30 kHz
5	Low frequency (LF)	30-300 kHz
6	Medium frequency (MF)	300 kHz- 3MHz
7	High frequency (HF)	3-300 MHz
8	Very high frequency (VHF)	30-300 MHz
9	Ultra high frequency (UHF)	300-3 GHz
10	Super high frequency (SHF)	3-30 GHz
11	Extremely high frequency (EHF)	30-300 GHz

Pertinent electromagnetic parameters governing the microwave heating: The loss tangent can be derived from material's complex permittivity. The real component of the permittivity is called the dielectric constant whilst the imaginary component is referred to as the loss factor. The ratio of the loss factor to the dielectric constant is the loss tangent. The complex dielectric constant is given by:

$$\varepsilon = \varepsilon' - j\varepsilon'' \qquad (4.6)$$

Where ε is the complex permittivity, ε' is the real part of dielectric constant; ε'' is the loss factor, and $\varepsilon''/\varepsilon' = \tan\delta$ is the loss tangent.

Knowledge of a material's dielectric properties enables the prediction of its ability to absorb energy when exposed to microwave radiation. The average power absorbed by a given volume of material when heated dielectrically is given by the equation:

$$P_{av} = \varpi \varepsilon_0 \varepsilon_{eff}" E_{rms}^2 V \tag{4.7}$$

Where P_{av} is the average power absorbed (W); ϖ is the angular frequency of the generator (rad/s); ε_0 is the permittivity of free space; $\varepsilon_{eff}"$ is the effective loss factor; E is the electric field strength (V/m); and V is the volume (m^3).

The effective loss factor $\varepsilon_{eff}"$ includes the effects of conductivity in addition to the losses due to polarization. It provides an adequate measure of total loss, since the mechanisms contributing to losses are usually difficult to isolate in most circumstances.

Another important factor in dielectric heating is the depth of penetration of the radiation because an even field distribution in a material is essential for the uniform heating. The properties that most strongly influence the penetration depth are the dielectric properties of the material. These may vary with the free space wavelength and frequency of the propagating wave. For low loss dielectrics such as plastics ($\varepsilon"\ll 1$) the penetration depth is given approximately by:

$$D_P = \frac{\lambda_0 \sqrt{\varepsilon'}}{2\pi \varepsilon_{eff}"} \tag{4.8}$$

Where D_P is the penetration depth; λ_0 is the free space wavelength; ε' is the dielectric constant; and $\varepsilon_{eff}"$ is the effective loss factor.

The penetration depth increases linearly with respect to the wavelength, and also increases as the loss factor decreases. Despite this, however, penetration is not influenced significantly when increasing frequencies are used because the loss factor also drops away maintaining a reasonable balance in the above equation

As the material is heated its moisture content will decrease. This leads to a decrease in the loss factor). It can be seen from equation (4.8) that the decrease in loss factor causes in the penetration depth of radiation.

Microwaves cause molecular motion by migration of ionic species and/or rotation of dipolar species. Microwave heating a material depends to a great extent on its "dissipation" factor, which is the ratio of dielectric loss or "loss" factor to dielectric constant of the material. The dielectric constant is a measure of the ability of the material to retard microwave energy as it passes through; the loss factor is a measure of the ability of the material to dissipate the energy. In other words "loss" factor represents the amount of input microwave energy that is lost in the material by being dissipated as heat. Therefore, a material with high "loss" factor is easily heated by microwave energy. In fact, ionic conduction and dipolar rotation are the two important mechanisms of the microwave energy loss (i.e., energy dissipation in the material). Non-homogeneous material (in terms of dielectric property) may not heat uniformly, that is, some parts of the materials heat faster than others. This phenomenon is often referred to as thermal runway.

Continuous temperature measurement during microwave irradiation is a major problem. Luxtron fluoroptic or accufiber can be employed to measure temperature up to 400°C but are too fragile for most industrial applications. An optical pyrometer and thermocouple can be employed to measure higher temperatures. Optical pyrometers, such as thermo-vision infrared camera, only records surface temperature, which is invariably much lower than the interior sample temperature. When a thermocouple (metallic probe) is employed for temperature measurements, arcing between the sample and the thermocouple can occur leading to temperature measurements, arcing between the sample and thermocouple can occur leading to failure in thermocouple performance. A recent development is the ultrasonic temperature probe, which covers temperature up to 1500 °C.

In summery, microwave heating is unique and offers a number of advantages over conventional heating such as:

- non-contact heating;
- energy transfer, not heat transfer;
- rapid heating;
- material selective heating;
- volumetric heating;
- quick start-up and stopping;
- heating starts from interior of the material body.

Some glossaries of microwave heating system are shown in Table 4.3[4].

Table 4.3. Some glossaries of microwave heating system

Applicator or cavity	A closed space where a material is exposed to microwaves for heating
Choke	Barriers placed at entrance and exit of the applicator to prevent leakage of microwaves.
Circulator	A three port ferrite device allowing transmission of energy in one direction but directing reflected energy into water load (dummy load) connected at the third port.
Coupling	The transfer of energy from one portion of a circuit to another.
Dielectric	It is a measure of a sample's ability to retard microwave energy as it passes through.
Dielectric loss or loss factor	It is a measure of a sample's ability to dissipate microwave energy.
Hertz (Hz)	1 Hz = 1 cycle/s.
Magnetron	An electronic tube for generating microwaves.
Single mode applicator	Dimension of applicator or cavity is comparable with the wave length of microwave.
Multimode applicator	An applicator dimension is large in relation to the wave legth of incident microwaves.

4.3. BASIC CONCEPTS OF MICROWAVE HEATING

As it was mentioned earlier, microwaves are electromagnetic waves having a frequency ranging from 300 MHz and 0.3 THz. Most of the existing apparatuses, however, operate between 400 MHz and 60 GHz, using well defined frequencies, allocated for industrial, Scientific and Medical (ISM) applications. Among them, the 2.45 GHz is widely used for heating applications, since it is allowed word-wide and it presents some advantages in terms of costs and penetration depth.

It was also mentioned earlier, that quantitative information regarding the microwave-material interaction can be deduced by measuring the dielectric properties of the material, in particular of the real and imaginary part of the relative complex permittivity, $\varepsilon = \varepsilon' - j\varepsilon''_{eff}$, where the term ε''_{eff} includes conduction losses, as well as dielectric losses. The relative permeability is not a constant and strictly depends on frequency and temperature. A different and more practical way to express the degree of interaction between microwaves and materials is given by two parameters; the power penetration depth (D_P)

and the power density dissipated in the material (P), as defined earlier in a simplified version as follows:

$$D_P = (\lambda_0 \sqrt{\varepsilon'})/2\pi\varepsilon'' \quad P = 2\pi f \varepsilon_0 \varepsilon''_{eff} E_{rms}^2 \quad (4.9)$$

where $\varepsilon = \varepsilon' - j\varepsilon''_{eff}$ is the complex permittivity of the material under treatment, λ_0 is the wavelength of the radiation, f is its frequency, $\varepsilon_0 = 8.854 \cdot 10^{-12}$ F/m is the permittivity of empty space and E_{rms} is the electric field strength inside the material itself. It should be noted that P and D_P can only give quantitative and often misleading information, especially when it is critical to determine the temperature profiles inside the material. Others are the variables involved, however from this two parameters can be deduced most of the peculiarities which make the microwave heating a unique process [5].

First of all, it can be noticed the existence of temperature profile inversion with respect to conventional heating techniques. The air in proximity of the materials during the heat treatment, in fact, is not a good microwave absorber so that it can be considered that the atmosphere surrounding the material is essentially at low temperature.

Viceversa, the material under treatment, interacting in a stronger way with the electromagnetic field, heats up and reaches higher temperature. The result is that, in most cases, the surface temperature of the sample is lower than inside the material itself. This effect is more pronounced for poor heat conducting materials.

Since the given formulation for D_P and P show a strong dependence upon the real and imaginary part of the material permittivity, for multiphase systems having components with quite different permittivity, it is expected a strong selectivity of the microwave heating process. Power, in fact, is transferred preferentially to some materials (with high ε''_{eff}) so that it can be possible to rise the temperature of just a single phase or component, or to spatially limit the heat treatment to the material, without involving the surrounding environment. This peculiarity can be particularly useful when treating composite materials.

The rapid variations of the permittivity as a function of temperature is responsible for a not always desirable phenomenon, the thermal runaway, that is to say the rapid and uncontrollable overheating of parts of the material under processing. Considering a low thermal conductivity material, whose permittivity increases as the temperature rises, in particular ε" increasing the temperature growing, it will be subject to gradient of temperature, being colder in the regions where heat is rapidly dissipated or the field strength is lower, and hotter in the remaining zones. These zones, presenting higher values of ε", and thus of P, will start absorbing microwaves more than the cold ones, further rising their temperature and consequently the local value of ε", strengthening the phenomenon.

Finally, dielectric heating is penetrating, depending on the operating wavelength, and permits to directly heat treat the surface and the core of the body, without waiting for the heat to reach the core of the sample by means of conduction, particularly time-taking for low thermal conductivity materials, like most polymers are. In these materials, the penetration depth is high, of the order of some tens of centimeters, thus facilitating the processing of large bodies, too [6].

4.4. HEAT AND MASS TRANSFER CLASSICAL EQUATIONS

The conservation of mass and energy for a textile material give the following equations:

$$\frac{\partial X}{\partial t} = D_n \nabla^2 X \tag{4.9}$$

$$C_{pn}\rho_n \frac{\partial T}{\partial t} = \nabla(K_n \nabla T) + Q(r,z,t) \tag{4.10}$$

where $n = 1$ and 2 refers to the inner and outer layer of material, and D is diffusivity (m^2/s); X the moisture content (kg/kg dry basis); k the thermal conductivity ($W/m\,K$); ρ the density (kg/m^3); C_p the heat capacity ($J/kg\,K$); and Q is the microwave source term (W/m^3).

The empirical model for calculating moisture diffusivity as a function of moisture and temperature;

$$D = \frac{1}{1+X} D_0 \exp\left[-\frac{E_0}{R}\left(\frac{1}{T} - \frac{1}{T_r}\right)\right] + \frac{1}{1+X} D_i \exp\left[-\frac{E_i}{R}\left(\frac{1}{T} - \frac{1}{T_r}\right)\right] \quad (4.11)$$

where D (m^2/s) is the moisture diffusivity; X the moisture content (kg/kg dry basis); T (°C) the material temperature; T_r a reference temperature, and R=0.0083143 kj/mol K is the ideal gas constant; $D_0(m^2/s)$ the diffusivity at moisture $X=0$ and temperature $T=T_r$; D_i (m^2/s) the diffusivity at moisture $X=\infty$ and temperature $T=T_r$; E_0 (kJ/mol) the activation energy of diffusion in dry material at $X=0$ and E_i

(kj/mol) is the activation energy of diffusion in wet material at $X=\infty$. The proposed model may uses the estimated parameters in Table 1.

Table 4.3. Numerical values for wool (based on data from various authors)

Diffusion coefficient of water vapor - 1st stage: $(1.04 + 68.20W_c - 1342.59W_c^2)10^{-14}$, $t < 540s$
Diffusion coefficient of water vapor - 2nd stage: $1.6164\{1 - \exp[-18.163\exp(-28.0W_c)]\}10^{-14}$, $t \geq 540s$
Diffusion coefficient in the air: $2.5e^{-5}$
Volumetric heat capacity of fiber: $373.3 + 4661.0W_c + 4.221T$
Thermal conductivity of fiber: $(38.49 - 0.720W_c + 0.113W_c^2 - 0.002W_c^3)10^{-3}$
Heat of sorption: $1602.5\exp(-11.72W_c) + 2522.0$
Porosity of fiber: 0.92
Density of fiber ; 1300 kg/m^3
Radius of fiber ; 1.03 e^{-5} m
Mass transfer coefficient: 0.137 m/s
Heat transfer coefficient: 99.4 $W/m^2 K$

- Initial conditions: At time t =0: $T = T_0(r,z)$, $X = X_0(r,z)$

- Boundary conditions: $\left.\dfrac{\partial X}{\partial t}\right|_{(r=0, H/2=0, t)} = 0$, $\left.\dfrac{\partial T}{\partial t}\right|_{(r=0, H/2=0, t)} = 0$

Figure 4.4 shows that there is an increase in drying rate because of the microwave power density.

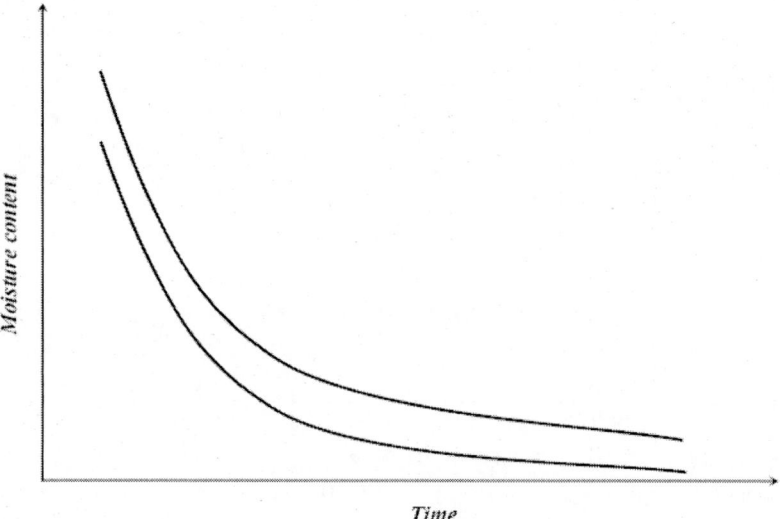

Figure 4.4. Comparison of conventional and microwave heating on average moisture content.

This can be attributed to the effect of microwave on moisture by rapidly increasing the moisture migration to the surface and increased evaporation. A comparison of these drying curves demonstrates improvement in drying times, under microwave heating. Nevertheless the results show significant improvement in average drying times over the conventional heating method.

4.5. HEAT AND MASS TRANSFER EXPONENTIAL MODEL

It has been recognized that microwaves could perform a useful function in textile drying in the leveling out of moisture profiles across a wet sample. This is not surprising because water is more reactive than any other material to dielectric heating so that water removal is accelerated. An exponential model presented here [7] can be used to describe the drying curves.

$$X = (a - X_{eq}).\exp(-b.t^d) + X_{eq} \qquad (4.12)$$

and its derivative form:

$$(-dX/dt) = b.d.t^{(d-1)}.X \qquad (4.13)$$

Parameters *a, b, d* can be determined by regression by the least square method. The quantities *b* and *d* vary with the experimental conditions and they are drying coefficients. X is the moisture content of the drying material, dX/dt is the drying rate and *t* is the drying time. Parameter *(a)* represents the initial moisture content.

From Figure 4.5 it is seen that the incident power strongly influenced the drying kinetics of a textile sample, reducing the drying time by raising the microwave heating power.

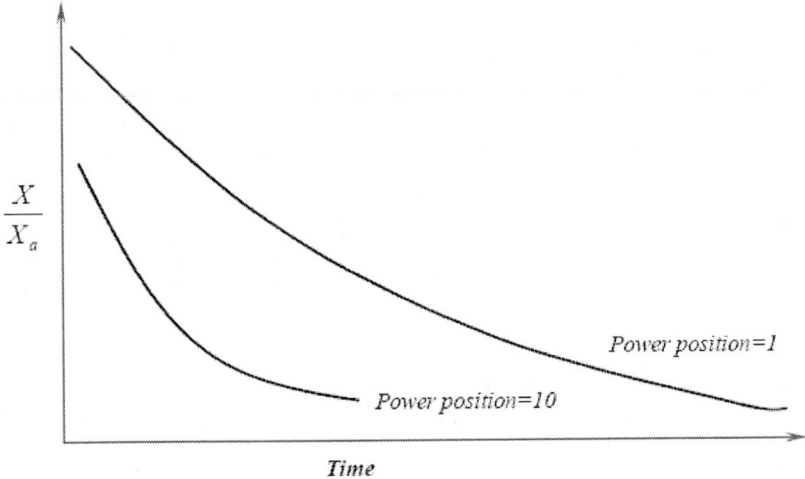

Figure 4.5. Normalized moisture content for two power of microwave heating.

4.6. COMBINED MICROWAVE AND CONVECTIVE DRYING OF TUFTED TEXTILE MATERIAL

It should be noted that because of the higher temperature and pressure gradients generated during combined microwave and convective drying, greater care must be taken not to damage the textile material to be dried, whilst still taking advantage of the increased drying rates provided by the microwave environment. To fully understand the heat and mass transfer phenomenon occurring within the material during combined microwave and convective drying, it is required to analyze the moisture, temperature and pressure distributions generated throughout the process

It was shown by Ilic and Turner [8] that a theory based on a continuum approach led to the following equations of motion governing the drying of a slab of material:

Total mass:

$$\frac{\partial}{\partial t}\left(\phi S_g \rho_g + \phi S_W \rho_W\right) + \nabla \cdot \left(\chi_g \rho_g V_g + \chi_W \rho_W V_W\right) = 0 \qquad (4.14)$$

Total liquid:

$$\frac{\partial}{\partial t}\left(\phi S_g \rho_{gv} + \phi S_W \rho_W\right) + \nabla \cdot \left(\chi_g \rho_{gv} V_{gv} + \chi_W \rho_W V_W\right) = 0 \qquad (4.15)$$

Here, S is the volume saturation, ϕ is the porosity, $\rho [kg\, m^{-3}]$ is the density of the fibers χ is the surface porosity, ϕ is the porosity

Total enthalpy:

$$\frac{\partial}{\partial t}\left(\phi S g \rho g v h g v + \phi S g \rho g a h g a + \phi S g \rho g a h g a + \phi S W \rho W h W + (1-\phi)\rho S h S - \phi \rho W \int_0^{S_W} \Delta h W (S) dS \right)$$
$$+ \nabla \cdot \left(\chi_g \rho_{gv} V_{gv} h_{gv} + \chi_g \rho_{ga} V_{ga} h_{ga} + \chi_W \rho_W V_W h_W \right)$$
$$= \nabla \cdot \left((K_g X_g + K_W \chi_W + K_S(1-\chi))\nabla T\right) + \phi \qquad (4.16)$$

Where ϕ is the internal microwave power dissipated per unit volume, K $[m^2]$ is permeability, and $h\ [Jkg^{-1}]$ is the averaged enthalpy. In equation (4.16) the effects of viscous dissipation and compression work have been omitted.

The equations (4.14, 4.15 and 4.16) are augmented with the usual thermodynamic relations and the following relations:

- Flux expressions are given as follows:

Gas flux:

$$\chi_g \rho_g V_g = -\frac{KK_g(S_W)\rho_g}{\mu_g(T)}\left[\nabla P_g - \rho_g g\right] \quad (4.16a)$$

Here, g $[ms^{-2}]$ is the gravitational constant and K_g is the relative permeability of gas.

Liquid flux:

$$\chi_W \rho_W V_W = -\frac{KK_W(S_W)\rho_W}{\mu_W(T)}\left[\nabla(P_g - P_C(S_W,T)) - \rho_W g\right] \quad (4.16b)$$

Where, K_W is the relative permeability of water, and μ $[Hm^{-1}]$ is the permeability of free spaces.

Vapor flux:

$$\chi_g \rho_{gv} V_{gv} = \chi_g \rho_{gv} V_g - \frac{\chi_g \rho_g D(T,P_g) M_a M_v}{M^2}\nabla\left(\frac{P_{qv}}{P_g}\right) \quad (4.16c)$$

Here, V $[ms^{-1}]$ is the averages velocity and M $[kgmol^{-1}]$ is the molar mass.

Air flux:

$$\chi_g \rho_{ga} V_{ga} = \chi_g \rho_g V_g - \chi_g \rho_g V_{gv} \quad (4.16d)$$

- Relative humidity (Kelvin effect):

$$\psi(S_W,T) = \frac{P_{gv}}{P_{gvs}(T)} = \exp\left(\frac{2\sigma(T)M_v}{r(S_W)\rho_W RT}\right) \quad (4.17)$$

where ψ is the relative humidity and $P_{gvs}(T)$ is the saturated vapour pressure given by the Clausius-Clapeyron equation.

- Differential heat of sorption:

$$\Delta h_W = R_v T^2 \frac{\partial(\ln\psi)}{\partial T} \quad (4.18)$$

- Enthalpy-Temperature relations:

$$h_{ga} = C_{pa}(T - T_R) \quad (4.19)$$

$$h_{gv} = h_{vap}^0 + C_{pv}(T - T_R) \tag{4.20}$$

$$h_W = C_{pW}(T - T_R) \tag{4.21}$$

$$h_s = C_{ps}(T - T_R) \tag{4.22}$$

The expressions for K_g, K_W are those given by Turner and Ilic [8], and μ_g, μ_W have had functional fits according to the data by Holman [9]. The diffusivity $D(T, P_g)$ given by Quintard and Puiggali [10] and the latent heat of evaporation given by,

$$h_{vap}(T) = h_{gv} - h_W \tag{4.23}$$

After some mathematical manipulations, the one-dimensional system of three non-linear coupled partial differential equations which model the drying process in a thermal equilibrium environment are given by:

$$a_{s1}\frac{\partial S_W}{\partial t} + a_{s2}\frac{\partial T}{\partial t} = \frac{\partial}{\partial Z}\left[K_{S1}\frac{\partial S_W}{\partial Z} + K_{T1}\frac{\partial T}{\partial Z} + K_{T1}\frac{\partial T}{\partial Z} + K_{P1}\frac{\partial P_g}{\partial Z} + K_{gr1}\right] \tag{4.24}$$

$$a_{T1}\frac{\partial S_W}{\partial t} + a_{T2}\frac{\partial T}{\partial t} = \frac{\partial}{\partial Z}\left(K_e\frac{\partial T}{\partial Z}\right) - \phi\rho_W h_{vap}\frac{\partial}{\partial Z}\left[K_S\frac{\partial S_W}{\partial Z} + K_T\frac{\partial T}{\partial Z} + K_P\frac{\partial P_g}{\partial Z}K_{gr}\right]$$

$$+ \left[\phi\rho_W C_{pW}\left(K_{S2}\frac{\partial S_W}{\partial Z} + K_{T2}\frac{\partial T}{\partial Z} + K_{P2}\frac{\partial P_g}{\partial Z} + K_{gr2}\right)\right]\frac{\partial T}{\partial Z} + \Phi(S_W, T) \tag{4.25}$$

$$a_{P1}\frac{\partial S_W}{\partial T} + a_{P2}\frac{\partial T}{\partial t} + a_{P3}\frac{\partial P_g}{\partial t} = \frac{\partial}{\partial Z}\left[K_S\frac{\partial S_W}{\partial Z} + K_T\frac{\partial T}{\partial Z} + K_{P3}\frac{\partial P_g}{\partial Z} + K_{gr3}\right] \tag{4.26}$$

The capacity coefficients a_{S1}, a_{T1}, a_{p1} and the kinetic coefficients K_{S1}, K_{T1}, K_{P1}, K_{gr1} all depend on the independent variables: Saturation S_W, Temperature T and total pressure P_g. The boundary conditions are written in one dimension as:

At z=0 (Drying surface):

$$K_{S1}\frac{\partial S_W}{\partial Z} + K_{T1}\frac{\partial T}{\partial Z} + K_{P1}\frac{\partial P_g}{\partial Z} + K_{gr1} = \frac{K_m M_V}{R\phi\rho_W}\left(\frac{P_{gV}}{T} - \frac{P_{gV0}}{T_0}\right)$$
(4.27a)

$$K_e\frac{\partial T}{\partial Z} - \phi\rho_W h_{Vap}\left(K_S\frac{\partial S_W}{\partial Z} + K_T\frac{\partial T}{\partial Z} + K_P\frac{\partial P_g}{\partial Z} + K_{gr}\right) = Q(T - T_0)$$
(4.27b)

$$P_g = P_o \quad (4.27c)$$

At z=L (Impermeable surface):

$$K_{S1}\frac{\partial S_W}{\partial Z} + K_{T1}\frac{\partial T}{\partial Z} + K_{P1}\frac{\partial P_g}{\partial Z} + K_{gr1} = 0$$
(4.28a)

$$K_e\frac{\partial T}{\partial Z} - \phi\rho_W h_{Vap}\left(K_S\frac{\partial S_W}{\partial Z} + K_T\frac{\partial T}{\partial Z} + K_P\frac{\partial P_g}{\partial Z} + K_{gr}\right) = 0$$
(4.28b)

$$(K_{S1} - K_S)\frac{\partial S_W}{\partial Z} + (K_{T1} - K_T)\frac{\partial T}{\partial Z} + (K_{P1} - K_{P3})\frac{\partial P_g}{\partial Z} + (K_{gr1} - K_{gr3}) = 0$$
(4.28c)

Initially:

$$T(z,0) = T_1 \quad (4.29a)$$

$$\frac{\partial P_c}{\partial Z} = -\rho_W g \quad (4.29c)$$

Figures 4.6 and 4.7 show compares convective dryings (with or without microwaves). Whilst for convective drying there are definite constant rate and falling rate periods, when microwaves are added the form of the curves change [2].

$$P_g(z,0) = P_0 \quad (4.29b)$$

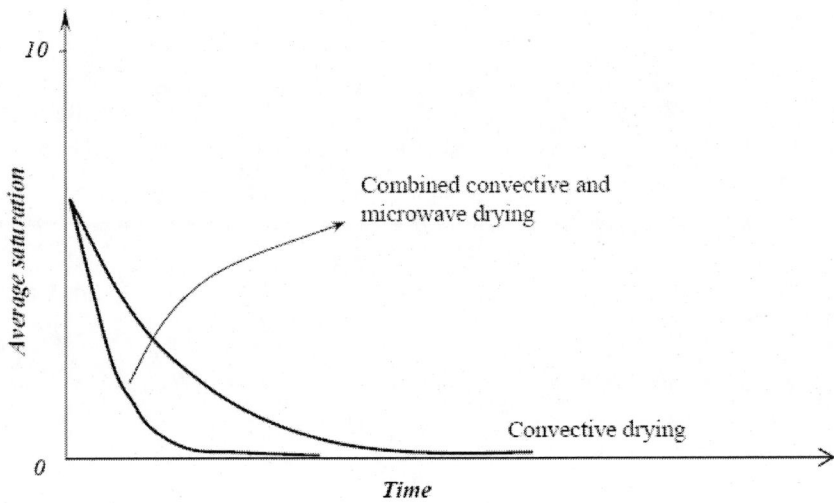

Figure 4.6. Average saturation profiles in time for drying with or without microwaves.

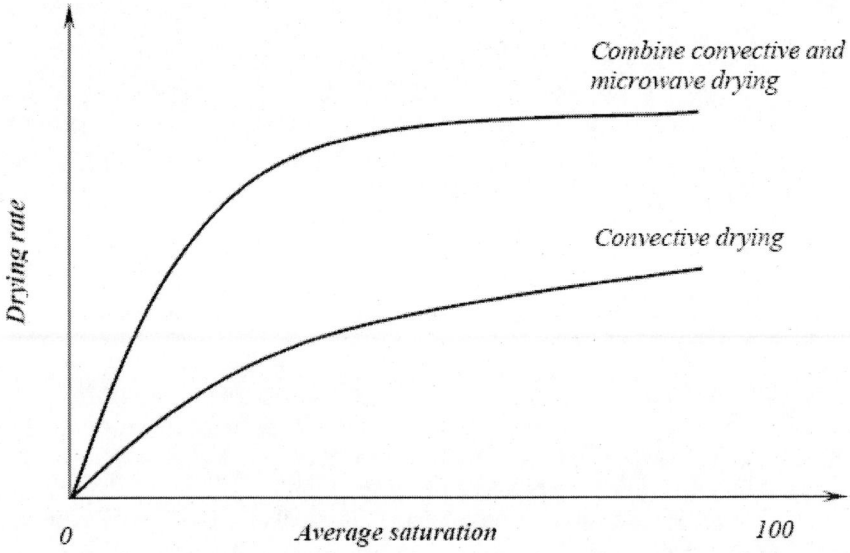

Figure 4.7. Drying rate curves corresponding to profiles plotted in Figure 4.6.

REFERENCES

[3] D.A.Jones, T.P. Lelyveld, S.D. Mavrofidis, S.V. Kingman, and N. J. Miles, Microwave heating application in environmental engineering, *Resources, Conservation and Recycling,* 34, 75-90, 2002.

[4] I.W. Turner and P.G. Jolly, Combined microwave and conventional drying of a porous material, Drying technology, 9(5), 1209-1269, 1991.

[5] T.V. Denend, Infrared predrying yields significant benefits, *American Dyestuff Reporter,* 45-51, December 1998.

[6] K. E. Haque, Microwave energy for mineral treatment processes, *Int. J. Miner. Process,* 57, 1-24, 1999.

[7] A.C. Metaxas and R.J. Meredith, *Industrial microwave heating,* Peter Peregrinus Ltd, London 1993, pp. 7-24.

[8] P. Veronesi, C. Leonelli, G. Pellacani and A. Boccaccini, Unique microstructure of glass-metal composites obtained by microwave assisted heat-treatment, *J. Thermal Anal. Cal.,* 72, 1141-1149, 2003.

[9] D. Skansi, Z. Bajza an A. Arapovic, Experimental evaluation of the microwave drying of leather, *J. of the society of Leather Thechnologies and chemists,* 79, 171-177.

[10] Ilic, M. and Turner, I.W.: Convective drying of a consolidated slab of wet porous material, *Int.J.Heat Mass Transfer,*, 32 (12),(1989).

[11] Holman, J. P., 1989, *Heat Transfer,* McGraw-Hill Book Company.

[12] Quintard M., and Puiggali J.R., 1986, Numerical modeling of transport processes during the drying of granular porous medium, *J. Heat and Technology,* 4, 2.

Chapter 5

HEAT FLOW AND CLOTHING COMFORT

Information on the transmission of water vapor by textiles fibers is desirable for better understanding of the problems of comfort, and data for design in special applications such as upholstery, footwear, immersion suits and other protective clothing, and wrapping or packaging, where high resistance to liquid water is desired, combined with considerable permeability of water vapour. Some of the issues of clothing comfort that are most readily understood involve the mechanisms by which clothing materials influence heat and moisture transfer from the skin to the environment. Heat transfer by convection, conduction and radiation and moisture transfer by vapor diffusion are the most important mechanisms in very cool or warm environments.

5.1. INTRODUCTION AND BACKGROUND

During physical activity the body provides cooling partly by producing insensible perspiration. If the water vapour cannot escape to the surrounding atmosphere the relative humidity of the microclimate inside the clothing increases causing a corresponding increased thermal conductivity of the insulating air, and the clothing becomes uncomfortable. In extreme cases hypothermia can result if the body loses heat more rapidly than it is able to produce it, for example when physical activity has stopped, causing a decrease in core temperature. If perspiration cannot evaporate and liquid sweat (sensible perspiration) is produced, the body is prevented from cooling at the same rate as heat is produced, for example during physical activity, and hyperthermia can result as the body core temperature increases. Table 5.1 shows heat energy produced by various activities and corresponding perspiration rates.

The ability of fabric to allow water vapour to penetrate is commonly known as breath-ability. This should more scientifically be referred to as water vapor permeability. Although perspiration rates and water vapor permeability are usually quoted in units of grams per day and grams per square meter per day, respectively, the maximum work rate can only be endured for a very short time. During rest, most surplus body heat is lost by conduction and radiation, whereas during physical activity, the dominant means of losing excess body heat is by evaporation of perspiration. It has been found that the length of time the body can endure arduous work decreases linearly with decrease in fabric water vapor permeability.

Table 5.1. Heat energy produced by various activities and corresponding perspiration rates [1]

Activity	Work rate (watts)	Perspiration rate (g/day)
Sleeping	60	2280
Sitting	100	3800
Gentle Walking	200	7600
Active Walking	300	11500
With light pack	400	15200
With heavy pack	500	19000
Mountain walking with heavy pack	600-800	22800-30400
Maximum work rate	1000-1200	38000-45600

It has also been shown that the maximum performance of a subject wearing clothing with a vapor barrier is some 60% less than that of a subject wearing the same clothing but without a vapor barrier. Even with two sets of clothing that exhibit a small variation in water vapor permeability, the differences in the wearer's performance are significant [1].

In an environment where body temperature cannot be regulated without a lot of sweating, we often try to get rid of heat from our body by turning on the air conditioning systems or moving into a conditioned room. Just after the change of the environment, we will feel "cool" or "comfortable". But the sweat accumulated in clothing evaporates gradually, until the heat loss from our body can be more than needed and at last we might feel "cold" or "uncomfortable". A review of clothing studies has shown that moisture collection in cold weather clothing, even after heavy exercise, seldom exceeds 10% by weight of added water [2]. One of the measurements are used to calculate values related to water vapor transmission properties is "water vapor resistance". This is the water vapor pressure difference across the two faces of the fabric divided by the heat flux per unit area, measured in square meters Pascal per watt. Some water vapor resistance data on different types of outwear fabrics are presented in Table 5.2. The measurement of water vapor resistance in thickness unit (mm) is the thickness of a still air layer having the same resistance as the fabric. The use of thickness units facilitates the calculations of resistance values for clothing assemblies comprising textile and air layers.

This observation is generally explained by noting that the major transfer mechanism from wet skin to underwear is one of distillation. An initial observation noting the surprisingly strong discomfort sensations associated with small amounts of water in the skin-clothing interface [3].

Table 5.2. Typical water vapor resistance(WVR) of fabrics[1]

Fabric, Outer (shell) material	WVR(mm still air)
Neoprene, rubber or PVC coated	1000-1200
Conventional PVC coated	300-400
Waxed cotton	1000+
Wool overcoating	6-13
Leather	7-8
Woven microfiber	3-5
Closely woven cotton	2-4
Ventile L28	3.5
Other Ventile	1-3
Two-layer PTFE laminates	2-3
Three-layer laminates (PTFE, polyester)	3-6
Microporous polyurethane (various types)	3-14

It has been confirmed in a number of studies in which either moisture from sweating or added moisture generates these clothing contact sensations. The procedures for these measurements [4] emphasize again that very little moisture is required to stimulate sensations of discomfort. Often 3% to 5% added moisture is ample to develop discomfort [5].

Simultaneous differential equations for the transfer of heat and moisture in porous medial under combined influence of gravity and gradients of temperature and moisture content were developed by D.A. De Vries [6]. These equations are a generalization of those drived by philip *et al.* [7]. Eckert and Faghri [8] have performed a general analysis of moisture migration in a slab of an unsaturated porous material for a condition where the temperature of one surface is suddenly increased to a higher value whereas the temperature of the other surface is maintained constant. Udell [9], [10] has derived a general, one-dimensional, steady-state model describing the heat and mass transfer within a homogeneous porous medium, saturated with a wetting liquid, its vapor and a non-condensible gas. The effects of gas diffusion, phase change, conduction, liquid and vapor transport, capillarity, and gravity are included. The analysis is based on a general thermodynamic description of the unique equilibrium states characteristics of liquid wetting porous media. Bouddor *et al.* [11] have provided a systematic, rigorous and unified treatment of the governing equations for simultaneous heat and mass transfer within a wide range of porous media.

Some work has also been done in the area of coupled diffusion of moisture and heat in hygroscopic textile materials. Gibson [12] has given a review of numerical modeling of convection, diffusion and phase changes in textiles. The paper summarizes current and past work aimed at utilizing CFD techniques for clothing applications. It was shown that water in a hygroscopic porous textile may exist in vapor or liquid form in the pore spaces. Phase changes associated with water include liquid evaporation/condensation in the pore spaces and sorption/desorption from polymer fibers. Additional factors such as swelling of solid polymer due to water and heat of sorption was incorporated into the appropriate conservation and transport equations. Nordon and David [13] have attempted to solve the non-linear differential equations which describe coupled diffusion of heat and mass (moisture) in hygroscopic textile materials. In addition to the diffusion equations, a rate equation was introduced describing the rate of exchange of moisture between the solid (textile fibers) and the gas phase. The predictions compared favorably with experimental observations on wool bales and wool fabrics [14]. Farnworth [15] has developed a simple model of combined heat and water vapor transport in clothing. Transport by forced convection was not included in this model.

Osczevski and Dolhan [16] and Farnworth *et al* [17] reported a strong dependency of water vapor resistance of hydrophilic membranes or coatings: the higher the relative humidity at the membrane, the lower the water vapour resistance (i.e., the higher the water vapor permeability or breathability).

In a temperature dependent experiment, Osczevski[18] placed a hydrophilic film on an ice block. Water vapor sublimating from the ice could diffuse only through the film and was collected by a desiccant. Osczevski measured mass transport through the film, and he found that water vapor resistance is an exponential function of temperature. In this experiment, water vapor permeability varnishes nearly completely with decreasing textile temperature. Because diffusion in hydrophilic materials is non-Fickian, he also derived from his results a theory of diffusion speed depending on activation energy, and he accounted for different relative humidity. Additionally, Gretton *et al* [19] reported an increase in the moisture vapor transmission rate of hydrophilic and microporous textiles when measuring with a heated dish instead of unheated dish. They interpreted their results by the increased motion of water vapor and polymer molecules, which they claimed would also work for micro-porous constructions.

Galbraith *et al* [20] compared cotton, water repellent cotton, and acrylic garments through wearing tests and concluded that the major factor causing discomfort was the excess amount of sweat remaining on the skin surface. Niwa [21] stated that the ability of fabrics to absorb liquid water (sweat) is more important than water vapor permeability in determining the comfort factor of fabrics.

Morooka and Niwa [22] postulated physiological factors related to the wearing comfort of fabrics as follows: sweating occurs whenever there is a tendency for the body temperature to rise, such as high temperature in the surrounding air and physical exercise, etc. If liquid water (sweat) cannot be dissipated quickly, the humidity of the air in the space in between the skin and the fabric that contacts with the skin rises. This increased humidity prevents rapid evaporation of liquid water on the skin and gives the body the sensation of "heat" that triggered the sweating in the first place. Consequently, the body responds with increased sweating to dissipate excess thermal energy. Thus a fabric's inability to remove liquid water seems to be the major factor causing uncomfortable feelings for the wearer.

Hollies [23] conducted wearer trails for shirts made of various fibers. They concluded that the largest factor that influenced wearing comfort was the ability of fibers to absorb water; regardless of weather fibers were synthetic or natural.

All of these studies indicate that the transient state phenomenon responding to the physiological demand to cause sweating is most relevant to comfort or discomfort associated with fabrics.

When work is performed in heavy clothing, evaporation of sweat from the skin to the environment is limited by layers of wet clothing and air. The magnitude of decrement in evaporative cooling is a function of the clothing's resistance to permeation of water vapor.

King and Cassie [24] conducted an experimental study on the rate of absorption of water vapor by wool fibers. They observed that, if a textile is immersed in a humid atmosphere, the time required for the fibers to come to equilibrium with this atmosphere is negligible compared with the time required for the dissipation of heat generated or absorbed when the regain changes. Gretton [25] investigated the effects of heat of sorption in the wool-water sorption system. They observed that the equilibrium value of the water content was directly determined by the humidity but that the rate of absorption and adsorption decreased as the heat-transfer efficiency decreased. Heat transfer was influenced by the mass of the sample, the packing density of the fiber assembly, and the geometry of the constituent fibers. Crank [26] pointed out that the water-vapor-uptake rate of wool is reduced by a rise in temperature that is due to the heat of sorption. The dynamic-water-vapor-sorption behavior of fabrics in the transient state will therefore not be the same as that of single fibers owing to the heat of sorption and the process to dissipate the heat released or absorbed.

Henry [27, 28] was who the first started theoretical investigation of this phenomenon. He proposed a system of differential equations to describe the coupled heat and moisture diffusion into bales of cotton. Two of the equations involve the conservation of mass and energy, and the third relates fiber moisture content with the moisture in the adjacent air. Since these equations are non-linear, Henry made a number of simplifying assumptions to derive an analytical solution.

In order to model the two-stage sorption process of wool fibers, David and Nordon [13] proposed three empirical expressions for a description of the dynamic relationship between fiber moisture content and the surrounding relative humidity. By incorporating several features omitted by Henry into the three equations, David and Nordon were able to solve the model numerically. Since their sorption mechanisms (i.e. sorption kinetics) of fibers were neglected, the constants in their sorption-rate equations had to be determined by comparing theoretical predictions with experimental results. Based on conservation equations, this global model consists of two differential coupled equations with variables for temperature and water concentration in air (C_a) and in the fibers of the textile (C_f), which is generally the water adsorbed by hygroscopic fibers. C_f is not in equilibrium with C_a, but an empirical relation between the adjustable parameters is assumed: the rate of sorption is a linear function of the difference between the actual C_f and the equilibrium value. The introduced coefficients are not directly linked to the physical properties of the clothes [29].

Farnworth [14] reported a numerical model describing the combined heat and water-vapor transport through clothing. The assumptions in the model did not allow for the complexity of the moisture-sorption isotherm and the sorption kinetics of fibers. Wehner *et al* [30] presented two mechanical models to simulate the interaction between moisture sorption by fibers and moisture flux through the void spaces of a fabric. In the first model, diffusion within the fiber was considered to be so rapid that the fiber moisture content was always in equilibrium with the adjacent air. In the second model, the sorption kinetics of the fiber were assumed to follow Fickian diffusion. In these models, the effect of heat of sorption and the complicated sorption behavior of the fibers were neglected.

Li and Holcombe [31] developed a two-stage model, which takes into account water-vapor-sorption kinetics of wool fibers and can be used to describe the coupled heat and moisture transfer in wool fabrics. The predictions from the model showed good agreement with experimental observations obtained from a sorption-cell experiment. More recently, Li and Luo [32] further improved the method of mathematical simulation of the coupled diffusion of the moisture and heat in wool fabric by using a direct numerical solution of the moisture-diffusion equation in the fibers with two sets of variable diffusion coefficients. These research publications were focused on fabrics made from one type of fiber. The features and differences in the physical mechanisms of coupled moisture and heat diffusion into fabrics made from different fibers have not been systematically investigated.

Holmer [33] compared the heat exchange and thermal insulation of two ensembles, one made from wool, the other from nylon, worn by subjects who exercised either lightly (dry condition) or strenuously (wet condition) for 60 minutes, then rested 60 minutes. He found that there was a significant difference in physiological and subjective responses between dry and wet conditions, but not between the two fiber types. Further, there was no significant difference between the ratings of temperature and humidity sensations for the wool and nylon garments. The wool garment picked up more water than the nylon garment (245 g versus 198 g) for the wet condition.

However, the wool fabric may have been slightly thicker than the nylon fabric, since it was reported to have a slightly greater thermal resistance and therefore hold more water [34].

Nielsen and Edrusick [29] evaluated the effect of five kinds of knit structures, all made from 100% polypropylene were evaluated. On subjects exercising for 40 minutes at 5°C followed by 20 minutes at rest, and then repeated. The thickest knit, a fleece, caused the greatest total sweat production, retained the most moisture, and wetted skin the most. They stated that the hydrophobic polypropylene prevented extensive sweat accumulation in the underwear (10 to 22%) causing the sweat to accumulate in the outer garments.

Bakkevig and Nielsen [35] repeated the protocol above, but used low and high work rates with three kinds of underwear (a polypropylene 1×1 knit, a wool 1×1 knit, and a fishnet polypropylene) worn under wool fleece covered by polyester/cotton outer garments. Total sweat production and evaporated sweat were the same for all three underwear fabrics, but where the sweat accumulated differed significantly. More sweat accumulated in the wool underwear than either polypropylene at both work rates. At the higher work rate, more sweat moved into the fleece layer from both kinds of polypropylene underwear than for the wool. Most likely for the 1×1 knits, the thicker wool underwear(1.95 mm) simply holds more water than the polypropylene underwear (1.41 mm) and based on outer layer-to layer wicking results, needs a greater volume of sweat to fill it pores before it starts to donate the excess to the layer above it.

Galbraith *et al* [34] conducted wear trails for shirts made of various fibers. They concluded that the largest factor that influenced wearing comfort was the ability of fibers to absorb water, regardless of whether fibers were synthetic or natural.

All of these studies indicate that the transient state phenomenon responding to the physiological demand to cause sweating is most relevant to comfort or discomfort associated with this general principle. It is important to point out that a highly water absorbing fabric placed in the first layer keeps the partial pressure of water vapor near the skin low, which helps dissipate water at the skin surface, although the water vapor transport rate is smaller than for non-absorbing fabrics.

In the other words, the dissipation of water by means of absorption by fabrics appears to be much more efficient way to keep the water vapor pressure near the skin low than dissipation by permeation through fabrics. Highly water absorbing fabrics raise the temperature of the air space near the skin. The temperature rise will further decrease relative humidity; however, the higher temperature may or may not desirable depending on environmental conditions [37].

In the literature, the emphasis has been placed on the correlation between sweating and discomfort associated with wearing fabrics. However, there is relatively less emphasis placed on the influence of changes in the surrounding conditions, that is, the influence of the seasons. Many comfort studies are conducted with a single layer of fabric at relatively warm and moderately humid conditions. Severe winter conditions, which mandate the use of layered fabrics, would necessitate totally different kinds of testing procedures. Consequently, it is necessary to distinguish the comfort factor and the survival factor, and to investigate these factors with different perspective[38].

The evaporation process is also influenced by the liquid transport process. When liquid water cannot diffuse into the fabric, it can only evaporate at the lower surface of the fabric. As the liquid diffuses into the fabric due to capillary action, evaporation can take place throughout the fabric [39].

Moreover, the heat transfer process has significant impact on the evaporation process in cotton fabrics but not in polyester fabrics. The process of moisture sorption is largely affected by water vapor diffusion and liquid water diffusion, but not by heat transfer. When there is liquid diffusion in the fabric, the moisture sorption of fibers is mainly determined by the liquid transport process, because the fiber surfaces are covered by liquid water quickly. Meanwhile, the water content distributions in the fibers are not significantly related to temperature distributions.

All moisture transport processes, on the other hand, affect heat transfer significantly. Evaporation and moisture sorption have a direct impact on heat transfer, which in turn is influenced by water vapor diffusion and liquid diffusion. The temperature rise during the transient period is caused by the balance of heat released during fiber moisture sorption and the heat absorbed during the evaporation process [40].

As a whole, a dry fabric exhibits three stages of transport behavior in responding to external humidity transients. The first stage is dominated by two fast processes: water vapor diffusion and liquid water diffusion in the air filling the inter-fiber void spaces, which can reach new steady states within fractions of seconds. During this period, water vapor diffuses into the fabric due to the concentration gradient across the two surfaces. Meanwhile, liquid water starts to flow out of the regions of higher liquid content to the dryer regions due to surface tension force [41].

The second stage features the moisture sorption of fibers, which is relatively slow and takes a few minutes to a few hours to complete. In this period, water sorption into the fibers takes place as the water vapor diffuses into the fabric, which increases the relative humidity at the surfaces of fibers. After liquid water diffuses into the fabric, the surfaces of the fibers are saturated due to the film of water on them, which again will enhance the sorption process. During these two transient stages, heat transfer is coupled with the four different forms of liquid transfer due to the heat released or absorbed during sorption/adsorption and evaporation/condensation. Sorption/adsorption and evaporation/condensation, in turn, are affected by the efficiency of the heat transfer. For instance, sorption and evaporation in thick cotton fabric take a longer time to reach steady states than in thin cotton fabrics.

Finally, the third stage is reached as a steady state, in which all four forms of moisture transport and the heat transfer process become steady, and the coupling effects among them become less significant. The distributions of temperature, water vapor concentration, fiber water content, and liquid volume fraction and evaporation rate become invariant in time. With the evaporation of liquid water at the upper surface of the fabrics, liquid water is drawn from capillaries to the upper surface.

5.2. Effective Thermal Conductivity

One way of expressing the insulating performance of a textile is to quote "effective thermal conductivity". Here the term "effective" refers to the fact that conductivity is calculated from the rate of heat flow per unit area of the fabric divided by the temperature gradient between opposite faces. It is not true condition, because heat transfer takes place by a combination of conduction through fibers and air and infrared radiation. If moisture is present, other mechanisms may be also involved. Research on the thermal resistance of apparel textiles [42-47], has established that the thermal resistance of a dry fabric or one containing very small amounts of water depends on its thickness, and to lesser extent on fabric construction and fiber conductivity. Indeed, measurements of effective thermal conductivity by standard steady-state methods show that differences between fabrics and mainly attributable to thickness. Despite these findings, consumers continue to regard wool as "warmer" than other fibers, and show preference for wearing wool garments in cold weather, particularly when light rain or sea spray is involved.

Meanwhile, the effective thermal conductivities of fabrics can be studied for varying regains. Regain is the mass of water present expressed as a percentage of the dry weight of the material. The effective thermal conductivities for porous acrylic, polypropylene, wool, and cotton is shown in Figure 5.1.

The curves indicate that changes in "effective thermal conductivity" with increasing regain are not linear, but can be explained in terms of water within the fibers of fabrics with regain.

Figure 5-2 presents the various phases diagrammatically. When fabrics containing water are subjected to a temperature gradient, three different modes of heat flow can be distinguished;

- the presence of condensed water
- vapor transport, and condensation.

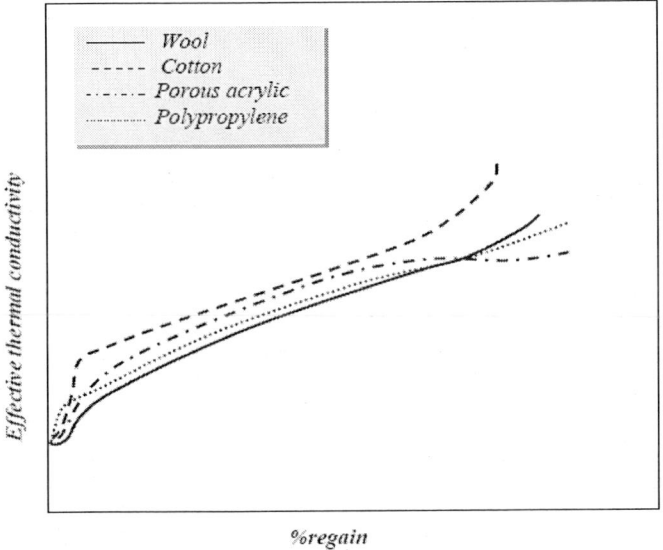

Figure 5.1. Comparison of effective thermal conductivities for porous acrylic, polypropylene, wool and cotton.

Figure 5.2.b. Cross sections of absorbent material at different regains.

Fiber sorption properties influence the heat and mass transfer up to the point when the rate of increased conductivity with regain is low in the curves, and then all fiber types behave similarly. Generally heat transfer increases with increasing regain, but in this initial regain the rise is most pronounces for the nonabsorbent polypropylene. The fiber with the lowest effective conductivity over the regain 0-200% regain is wool, an effect that is especially pronounced in the region of low regains from zero to saturation. This is mostly influenced by fiber sorption properties. Low regains are most common in real wear situation. This is mostly influenced by fiber sorption properties. Low regains are most common in real wear situations.

This explains the popular association between wool and warmth in situations such as yachting, where the garment will very likely become wet. Cotton fabric has the highest effective thermal conductivity for almost the whole regain.

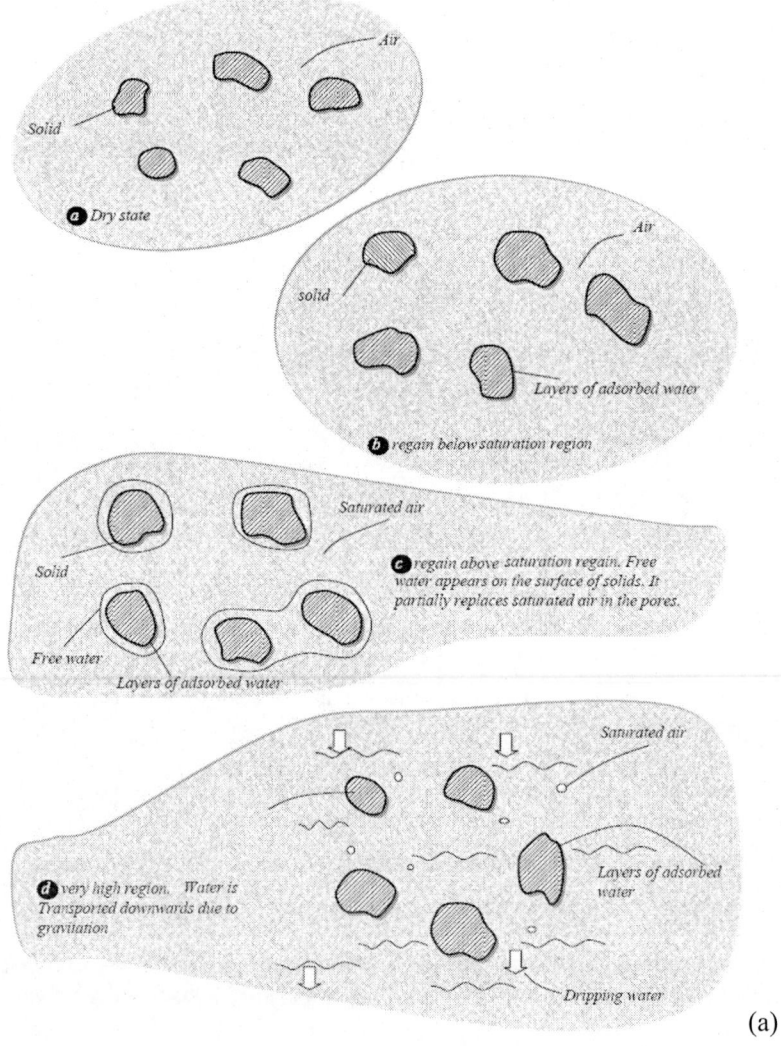

Figure 5.2.a. Cross sections of nonabsorbent material at different regains.

Heat Flow and Clothing Comfort

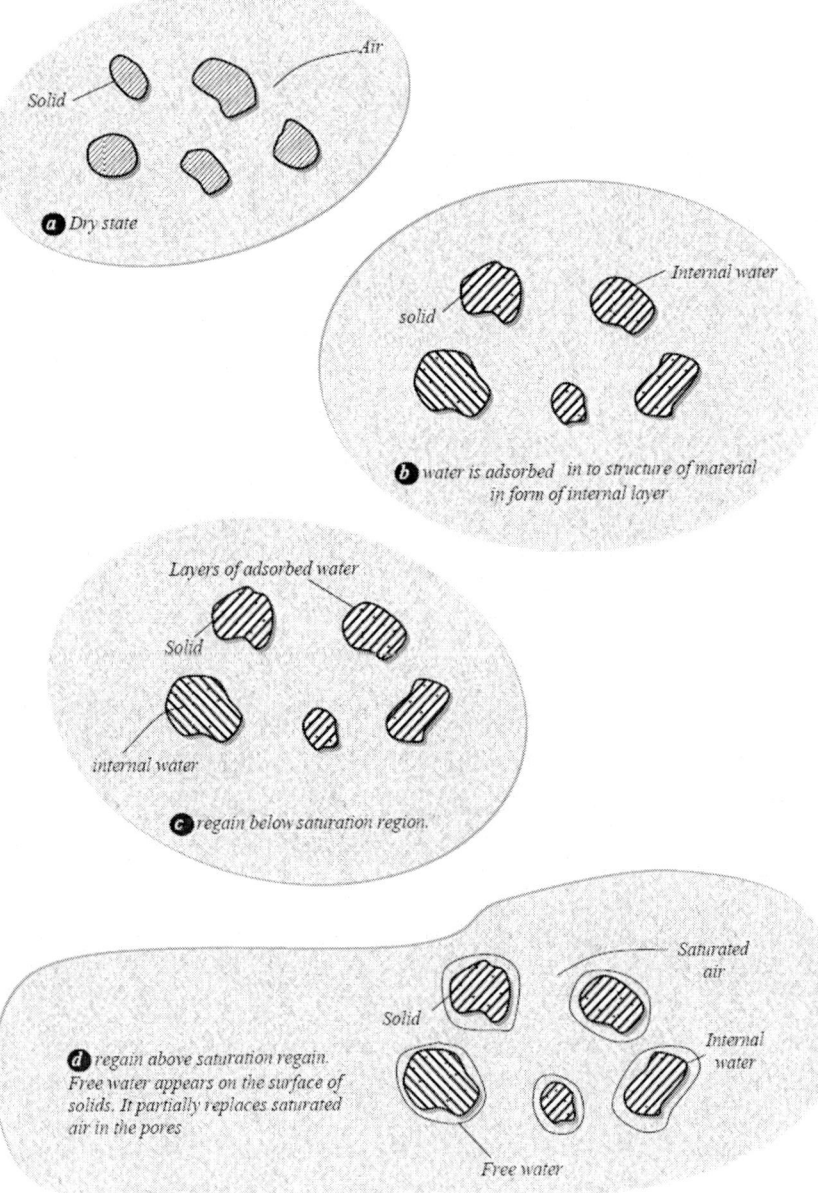

Figure 5.2.b. (Continued)

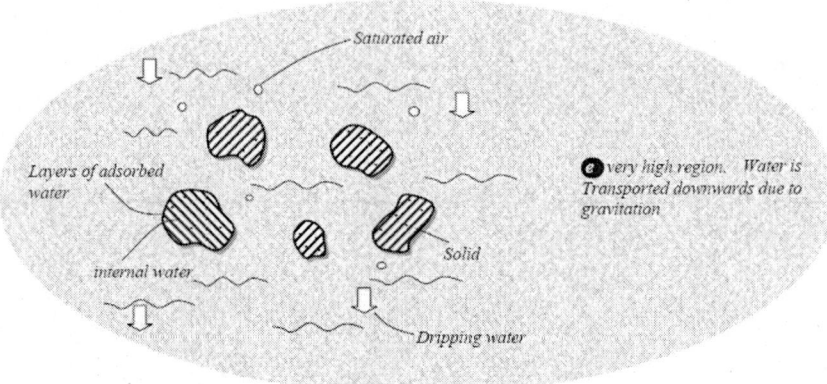

Figure 5.2.b. Cross sections of absorbent material at different regains.

5.3. TRANSPORT PHENOMENA FOR SWEAT

Fabrics to protect human body are, in most cases used under non-equilibrium conditions; therefore, characteristics of fabrics under non-isothermal and non-equilibrium conditions are important in evaluating overall performance. Furthermore, in colder environments, layered fabrics rather than a single fabric are used in most cases. Under such conditions, the two most important characteristics of fabrics are water vapor and heat transport. However, the water vapor transport may not influenced significantly by surface characteristics-the hydrophilic or hydrophobic nature of fabrics. On the other hand, when liquid water contacted a fabric, such as in the case of sweating, the surface wet-ability of fabric play a dominant role in determining the water vapor transport through layered fabrics [48-50].

In such a case, the wicking characteristics, which determines how quickly and how widely liquid water spreads out laterally on the surface of or within the matrix of the fabric, determines the overall water vapor transport rate through the layered fabrics.

It should be noted that the overall water vapor and heat transport characteristics of a fabric should depend on other factors such as the water vapor absorbability of the fibers, the porosity, density, and thickness of the fabric, *etc*[51-53].

Moreover, transport phenomena for the sweat case are much more complicated than the water vapour case because wetting of the surface by liquid water precedes water wetting of the surface by liquid water precedes water vapour transmission. Note that there is an important difference in water absorbing characterizing of wool and cotton, although both fibers have relatively high water vapor absorption rates. Because of the hydrophobic surface of wool fibers, a liquid droplet in contact with a wool fabric does not spread out laterally within a fabric layer. The water vapor transport rate, in the sweat case, can be indicated by the size of liquid water spread out on the surface or within a fabric matrix[54,55].

Moreover, the term "breathable" implies that the fabric is actively ventilated. This is not the case. Breathable fabrics passively allow water vapour to diffuse through them yet still prevent the penetration of liquid water. Production of water vapour by the skin is essential for maintenance of body temperature. The normal body core temperature is 37°C, and skin temperature is between 33 and 35°C, depending on conditions. If the core temperature goes beyond critical limits of about 24°C and 45°C then death results. The narrower limits of 34°C and 42°C can cause adverse effects such as disorientation and convulsions. If the sufferer is engaged in a hazardous pastime or occupation then this could have disastrous consequences.

5.4. FACTORS INFLUENCING THE COMFORT ASSOCIATED WITH WEARING FABRICS

As it was mentioned earlier, if liquid water (sweat (sweat)) cannot be dissipated quickly, the humidity of the air in the space in between the skin and the fabric that contacts with the skin rises. This increased humidity prevents rapid evaporation of liquid water on the skin and gives the body the sensation of "heat" that triggered the sweating in the first place. Consequently, the body responds with increased sweating to dissipate excess thermal energy. Thus a fabric's inability to remove liquid water seems to be the major factor causing uncomfortable feeling for the wearer.

Hollies *et al.* [38] conducted wearer trails for shirts made of various fibers.

They concluded that the largest factor that influence wearing comfort was the ability of fibers to absorb water, regardless of whether fibers were synthetic or natural. All of these studies indicate that the transient state phenomenon responding to the physiological demand to cause sweating is most relevant to comfort or discomfort associated with fabrics. It is important to point out that a highly water absorbing fabric placed in the first layer keeps the partial pressure of water vapor near the skin low, which helps to dissipate water at the skin surface, although the vapor transport rate is smaller than for non-absorbing fabrics. In other words, the dissipation of water by means of absorption by fabrics appears to be much more efficient way to keep water vapor pressure near the skin low than dissipation by permeation through fabrics. Highly water absorbing fabrics raise the temperature of the air space near the skin. The temperature rise will further decrease relative humidity; however, the higher temperature may or may not be desirable depending on environmental conditions.

5.5. INTERACTION OF MOISTURE WITH FABRICS

To stay warm and dry while active outdoors in winter has always been a challenge. In the worst case, an individual exercises strenuously, sweats profusely, and then rests. During exercise, liquid water accumulates on the skin and starts to wet the clothing layers above skin. Some of the sweat evaporates from both the skin and the clothing. Depending on the temperature and humidity gradient across the clothing, the water vapor either leaves the clothing or condenses and freezes somewhere in its outer layers.

When one stops exercising and begins to rest, active sweating soon ceases. This allows the skin and clothing layers eventually dry. During this time, however, the heat loss from body can be considerable. Heat is taken from the body to evaporate the sweat, both that on the skin and that in the clothing. The heat flow from the skin through the clothing can be considerably greater when the clothing is very wet, since water decreases clothing's thermal insulation. This post-exercise chill can be exceedingly comfortable and can lead to dangerous hypothermia.

A dry layer next to the skin is more comfortable than a wet one. If one can wear clothing next to the skin that does not pick up any moisture, but rather passes it through to a layer away from the skin, heat loss at rest will be reduced. For such reasons, synthetic fibers have gained popularity with winter enthusiasts such as hikers and skiers.

Advertising the popular press would have us believe that synthetic materials pick up very little moisture, dry quickly, and so leave the wearer warm and dry. In contrast, warnings are given against wearing cotton or wool next to skin, since these fibers absorb sweat and so "lower body temperature". A further property credited to synthetics, in particular polypropylene, is that they wick water away from the skin, leaving one dry and comfortable.

In the early fifties, when synthetic fibers such as nylon and the acrylics were first coming onto the consumer market, Fourt et al. [56] and Coplan [57] compared the water absorption and drying properties of these "miracle" fibers with those of conventional wool and cotton. Forty-five years latter, the water absorption and drying properties of synthetics were compared with natural fibers and it was found that all fabrics pick up water, and the time they take to dry is proportional to the amount of water they initially pick up [58,59].

It was also found that properties relevant to clothing on an exercising person, that is, the energy required to evaporated water from under and through a dry fabric or to dry a wet fabric and layer-to-layer wicking [60].

Holmer [61] compared the heat exchange and thermal insulation of two ensembles, one made from wool, the other from nylon, worn by subjects who exercised either lightly (dry condition) or strenuously (wet condition) for 60 minutes, then rested 60 minutes.

He found that there was a significant difference in the physiological and subjective responses between dry and wet conditions, but not between the two fiber types. Further, there was no significant difference between the ratings of temperature and humidity sensations for the wool and nylon garments. The wool garment picked up more water than the nylon garment (245 g versus 198 g) for the wet condition. However, the wool fabric may have been slightly thicker than the nylon fabric, since it was reported to have a slightly greater thermal resistance and would therefore, holding more water.

5.6. MOISTURE TRANSFER IN TEXTILES

In nude man any increase of sweating is immediately accompanied by an increase in heat loss due to evaporation. Similarly any decrease in sweating is immediately accompanied by a decrease in heat loss. Thus, nude man has a control of his heat loss which has no appreciable time lag. This is shown diagrammatically in Figure 5.3.

In this figure time is plotted as abscissa and rate of heat produced or lost as ordinate. To maintain perfect heat balance and a constant temperature, heat loss should equal heat production so that the heat production and heat loss curves should be the same. Suppose a man is initially at rest with a low heat production and a like heat loss as represented by the solid line in period A. When he exercises and produce more heat, the heat loss should rise as represented by the solid line in period B. Again, when he returns to the resting condition, period C, heat loss should return to the solid base line.

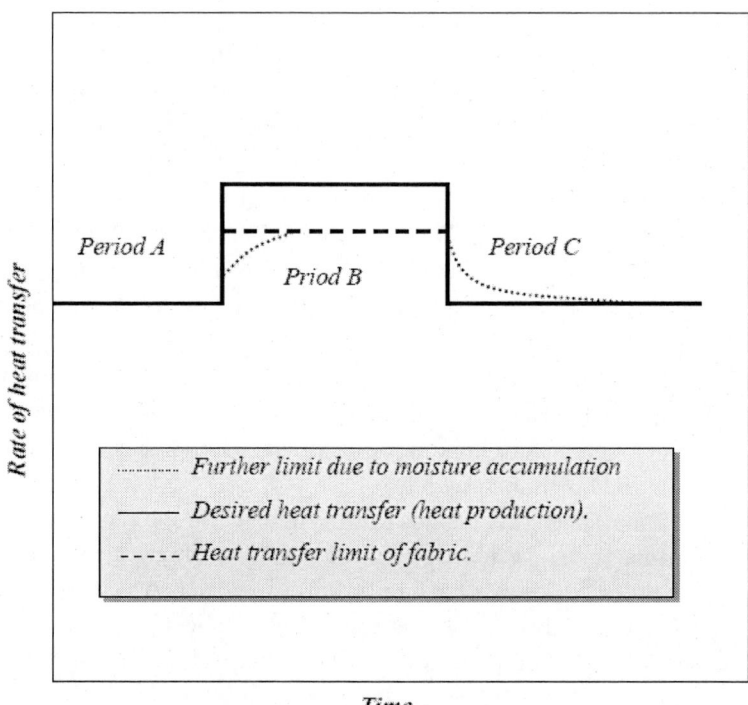

Figure 5.3. Rate of heat transfer versus time.

If sweating is the mechanism bringing about increasing heat loss but evaporation is limited, the increased heat loss might only be sufficient to match an increased heat production represented by the dashed line in period B. The position of this dashed line will depend on the equilibrium vapor transfer characteristics of the clothing. If, however, the hypothetical man is clothed in absorbent clothing, some of the sweat initially evaporated at the skin at the start of the exercise period will be absorbed by the clothing and its heat of absorption will appear in the clothing as sensible heat. This source of sensible heat will temporarily reduce the heat loss so that it follows the dotted line. Eventually a new equilibrium moisture content will be established and the dotted and dashed lines will coincide. When exercise and sweating stop, period C, moisture accumulated in the clothing will be desorbed or evaporated and tend to cool the clothing and the man wearing it. Thus, there is a time lag, and the heat loss curve will tend to follow the dotted curve during the after-exercise period. Since in Figure 5.3 heat loss per unit time is plotted against time, the area between the dotted line and the solid line represents an amount of heat, as distinguished from rate of heat loss, which can be regarded as a quantitative value of after exercise chill [62].

It should be noted that the moisture contained in the clothing need not be only that which is collected by absorption. It is also possible in cold damp or extreme cold environments that sweat which is evaporated at the skin will re-condense when it reaches colder layers of clothing. Alternatively the sweat rate may be so high that some of it will not evaporate from the skin. In nude man this drips off, but in clothed man it is blotted up by clothing to evaporate after sweating ceases.

Meanwhile, measurements of water vapor permeability of woven fabrics have indicated that in the lower ranges of fabric density, the main path of water vapor transfer is through the air spaces between fibers and yarns. This covers the densities characteristic of most apparel fabrics made from staple fibers, although filament yarn fabrics may be woven to higher densities in which the kind of fiber itself in the passage of water vapor , it is necessary to account for the water vapor passage through air spaces.

5.7. WATER VAPOR SORPTION MECHANISM IN FABRICS

In 1393, Henry [63] proposed a mathematical model for describing heat and moisture
Transfer in fabric, as shown in Equations 5.1 and 5.2, and the further analyzed the model in 1948 [64];

$$\varepsilon \frac{\partial C_a}{\partial t} + (1-\varepsilon)\frac{\partial C_f}{\partial t} = \frac{D_a \varepsilon}{\tau} \frac{\partial^2 C_a}{\partial x^2} \tag{5.1}$$

$$C_v \frac{\partial T}{\partial t} - \lambda \frac{\partial C_f}{\partial t} = K \frac{\partial^2 T}{\partial x^2} \tag{5.2}$$

In these equations, both C_v and λ are functions of the concentration of water absorbed by the fibers. Most textile fibers have very small diameters and very large surface/volume ratios. The assumption in the second equation of instantaneous thermal equilibrium between the fibers and the gas in the inter-fiber space does not therefore lead to appreciable error. The two equations in the model are not linear and contain three unknown, *i.e.*, C_f, T, and C_a . A third equation should be established appropriately in order to solve the model. Henry [63, 64] derived a third equation to obtain an analytical solution by assuming that C_f is linearly dependent on T and C_a, and that fibers reach equilibrium with adjacent air instantaneously. Considering the two-stage sorption process of wool, David and Nordon [65] proposed an exponential relationship to describe the rate of water content change in the fibers, as shown in equations (5.3) and (5.4);

$$\frac{1}{\varepsilon}\frac{\partial C_f}{\partial t}(H_a - H_f)\gamma \qquad (5.3)$$

where

$$\gamma = k_1(1 - \exp(k_2|H_a - H_f|)) \qquad (5.4)$$

and k_1 and k_2 are adjustable parameters that are evaluated by comparing the prediction of the model and measured moisture content of the fabric.

Farnworth [66] reported a numerical model describing combined heat and water vapour transport through clothing.

The assumptions in his model do not allow for the complexity of the moisture sorption isotherm and the sorption kinetics of fibers. Wehner et al [67] presented two mathematical models to simulate the interaction between moisture sorption by fiber and moisture flux through the air spaces of a fabric. In the first model, they considered diffusion within the fiber to be so rapid that the fiber moisture content is always in equilibrium with the adjacent air. In the second model, they assumed that the sorption kinetics of the fiber follows Fickian diffusion. Their model neglected the effect of heat of sorption behavior of the fiber. Li and Luo [32] developed a new sorption equation that takes into account the two-stage sorption kinetics of wool fibers, and incorporated this with more realistic boundary conditions to simulate the sorption behavior of wool fabrics. They assumed that water vapor uptake rate of fiber consists of a two components associated with the two stages of sorption identified by Downes and Mackay [68] and described by Watt [69].

The first stage is represented by Fickian diffusion with a constant coefficient. Second-stage sorption is much slower than the first and follows an exponential relationship. The relative contributions of the two stages to the total uptake vary with the sorption stage and the initial regain of the fibers. Thus, the sorption rate equation can be written as;

$$\frac{\partial C_f}{\partial t}(1-p)R_1 + pR_2 \qquad (5.5)$$

where R_1 is the first-stage sorption rate, R_2 is the second-stage rate sorption rate, and p is a proportional of uptake in the second stage. Equation (5.5) assumes that the sorption rate is a linear average of the first and second sorption rates. The first-stage sorption rate R_1 can be derived using Crank's truncated solution [26].

This may lead to a corresponding algorithm that needs a strict time striction and hence long computation times.

The second-stage sorption rate R_2, which relates local temperature, humidity, and the sorption history of the fabric, is assumed to have the following form;

$$R_2(x,t) = s_1 sign(H_a(x,t) - H_a(x,t) - H_f(x,t))$$
$$\times \exp\left(\frac{s_2}{|H_a(x,t) - H_f(x,t)|}\right) \qquad (5.6)$$

where s_1 and s_2 are constants. No values for s_1 and s_2 have been reported in the literature for any textile fibers. This is also an empirical equation that has an unclear physical meaning, which makes it inconvenient to predict and simulate heat and moisture transport in a fabric. These equations were improved substantially by Li and Luo [32]. The numerical values and approximate relationships they used are listed in Table 5.3. They assumed that moisture sorption by a wool fiber can be generally described by a uniform diffusion equation for both stages of sorption;

$$\frac{\partial C_f(x,r,t)}{\partial t} = \frac{1}{r}\frac{\partial}{\partial r}\left(rD_f(x,t)\frac{\partial C_f(x,r,t)}{\partial r}\right) \qquad (5.7)$$

$$C_{fs}(x,R_f,t) = f(H_a(x,t),T(x,t))$$

where $D_f(x,t)$ are the diffusion coefficients that have different presentations at different stages of sorption, and x is the coordinate of a fiber in the given fabric. The boundary condition is determined by the relative humidity of the air surrounding a fiber at x. In wool fabric, $D_f(x,t)$ is a function of $W_c(x,t)$, which depends on the sorption time and the fiber location.

Table 5.3. Numerical values of wool and physical properties

Parameters	Initial values	Mathematical relationship
Thermal conductivity of fabric (KJ/m.K)	$3.8493e^{-2}$	$(38.493 - 0.72W_c + 0.113W_c^2 - 0.002W_c^3)10^{-3}$
Volumetric heat capacity of fabric $(kJ/m^3.K)$	1609.7	$373.3 + 4661\,W_c + 4.221\,T$
Diffusion coefficient of fiber (m^2/s)	$2.4435\,e^{-14}$	$1.0637\,\arctan(1541.1933)$ $(3600/t^2)10^{-14}$
Diffusion coefficient of water vapor in fabric (m^2/s)	$1.91\,e^{-5}$	—
Heat of sorption or adsorption of water by fibers (KJ/Kg)	4124.5	$1602.5\exp(-11.72\,W_c) + 2522$
Porosity of fabric	0.925	—
Density of fabric (Kg/m^3)	1330	—
Radius of wool fiber (m)	$1.04e^{-5}$	—
Mass transfer coefficient (m/s)	0.137	—
Heat transfer coefficient $(W/m^2.K)$	99.4	—

W_c = Water content of the fibers in the fabric.

5.8. MODELING

The fabric model simulates the transport of a liquid and vapor-phase fluid that can undergo phase change (e.g., water) and an inert gas (air) in a textile layer. Several new models and capabilities were added to a standard commercial CFD code (FLUENT Version 6.0, Fluent Inc., Lebanon, NH) [47]. These capabilities include:

- Vapor phase transport (variable permeability).
- Liquid phase transport (wicking).
- Fabric property dependence on moisture content.
- Vapor/liquid phase change (evaporation/condensation).
- Sorption to fabric fibers.

In the fabric, transport equations are derived for mass, momentum, and energy in the gas and liquid phases by volume-averaging techniques. Definitions for intrinsic phase average, global phase average, and spatial average for porous media are those given by Whitaker [48]. Since the fabric porosity is not constant due to changing amounts of liquid and bound water, the source term for each transport equation includes quantities that arise due to the variable porosity. These equations are summarized in general form below [47].

Gas phase continuity equation:

$$\frac{\partial}{\partial t}\left((1-\varepsilon_{ds})\rho_\gamma\right) + \nabla \cdot (\rho_\gamma v_\gamma) = S_\gamma \quad (5.8)$$

$$S_y = m'_{sv} + m'_{lv} + \frac{\partial}{\partial t}\left((\varepsilon_{bl} + \varepsilon_\beta)\rho_\gamma\right) \quad (5.9)$$

Vapor continuity equation:

$$\frac{\partial}{\partial t}\left((1-\varepsilon_{ds})\rho_\gamma m_v\right) + \nabla \cdot (\rho_\gamma m_v v_\gamma) = \nabla \cdot \{\rho_\gamma D_{eff} \nabla(m_v)\} + S_v \quad (5.10)$$

$$S_v = m'''_{sv} + m'''_{lv} + \frac{\partial}{\partial t}\{(\varepsilon_{bl} + \varepsilon_\beta)\rho_\gamma m_v\} \quad (5.11)$$

$$m_v = \frac{\rho_v}{\rho_\gamma} \tag{5.12}$$

Gas phase momentum equation:

$$\frac{\partial}{\partial t}(\rho_\gamma v_\gamma) + \nabla \cdot (\rho_\gamma v_\gamma) = \nabla \cdot \{\mu_\gamma \nabla (v_\gamma)\} - \nabla p_\gamma + S_\gamma \tag{5.13}$$

$$S_\gamma = -v_\gamma \frac{\mu_\gamma}{K_\gamma k_\gamma} \tag{5.14}$$

Liquid transport:

$$\frac{\partial}{\partial t}\left[(1-\varepsilon_{ds})\rho_\gamma s\right] = \nabla \cdot \left(-\frac{k_\beta K_\beta}{\mu_\beta} \frac{\partial P_c}{\partial s}\right)\nabla s + S_l \tag{5.15}$$

$$S_l = -\nabla \cdot \left(\frac{k_\beta K_\beta}{\mu_\beta} \rho_\beta g\right) - \frac{m'''_{ls}}{\rho_\beta} - \frac{m'''_{lv}}{\rho_\beta} + \frac{\partial}{\partial t}[(\varepsilon_{bl})s] +$$

$$\frac{\partial}{\partial t}\left[(1-\varepsilon_{ds})(\rho_\gamma - 1)s\right] \tag{5.16}$$

$$\frac{\partial}{\partial t}\left[(1-\varepsilon_{ds})\rho_\gamma h_\gamma + \varepsilon_{ds}\rho_{ds}h_{ds}\right] + \nabla \cdot (\rho_\gamma v_\gamma h_\gamma) + \nabla \cdot (\rho_\beta v_\beta (h_v - \Delta h_v))$$

$$= \nabla \cdot [k_{eff} \nabla(T)] + \nabla \cdot (-h_a J_a - h_v J_v) + S_T \tag{5.17}$$

$$S_T = \frac{\partial}{\partial t}\left[(\varepsilon_\beta + \varepsilon_{bl})\rho_\gamma h_\gamma - \varepsilon_\beta \rho_\beta (h_v - \Delta h_v) - \varepsilon_{bl}\rho_\beta (h_v - \Delta h_v - Q)\right] \tag{5.18}$$

$$v_\beta = \frac{k_\beta K_\beta}{\mu_\beta}\left[\frac{\partial P_c}{\partial s}\nabla s + \rho_\beta g\right] \tag{5.19}$$

In summery, due to the intensive body activity, the wearer perspires and the cloth worn next to skin will get wet. These moisture fabrics reduce the body heat and make the wearer to become tired. So the cloth worn next to the skin should assist for the moisture release quickly to the atmosphere. The fabric worn next to the skin should have two important properties. The initial and fore most property is to evaporate the perspiration from the skin surface and the second property is to transfer the moisture to the atmosphere and make the wearer to feel comfort. Diffusion and wicking are the two ways by which the moisture is transferred to the atmosphere. These two are mostly governed by the fiber type and fabric stricture. The air flow through the fabric makes the moisture to evaporate to the atmosphere. The capillary path plays a vital role in the transfer of moisture and this depends on the wicking behavior of the fabric. In the development of protective clothing and other textiles, modeling offers a powerful companion to experiments and testing.

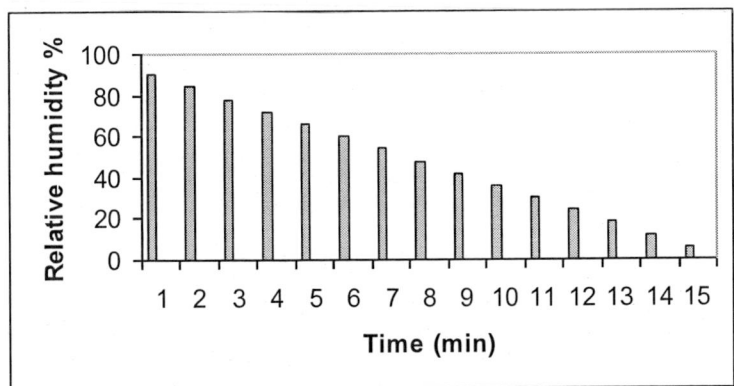

Figure 5.4. Relative hygrometry of wool fiber during adsorption.

It should be noted that condensation occurs when the vapor density of the steam is higher that its saturation vapor density. The condensation rate is proportional to the vapor density difference between that in the gas phase and that at the condensing surface. The relative hygrometry is quite different due to the action (water vapor pressure) of the airs on each side of the fabric (Figure 5.4). Sorption and adsorption have not opposite kinetics, the former faster in temperature and in charge of humidity during the first minutes, the latter more complete in discharge of humidity after a long time (figure 5.5). Figure 5.6 shows experimental results of hygrometry for the internal and external surfaces of wool fabric(between skin and fabric). Meanwhile, the internal gap has a considerable effect on the moisture transmission rate. The internal air gap has been identified as being a source of potential errors in most experimental works due to its changing resistance (Figure 5.7). Figure 5.8 shows the effect of thickness on the amount of water can be held in a fabric.

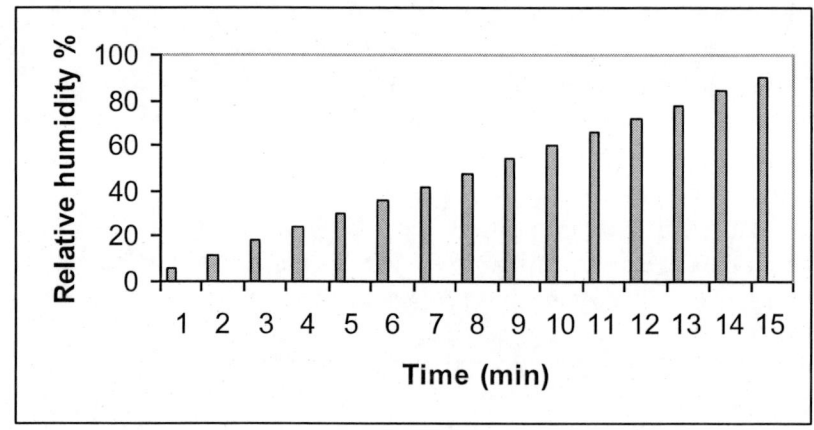

Figure 5.5. Relative hygrometry of wool fiber during sorption.

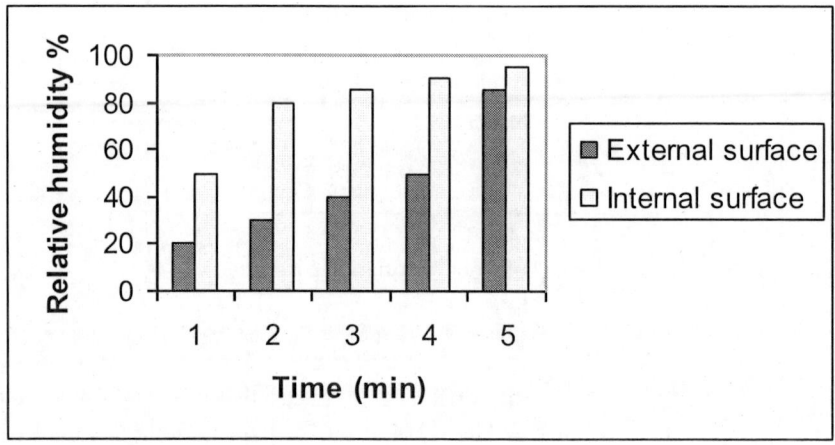

Figure 5.6. Experimental results of hygrometry for the internal and external surfaces of wool fabric(between skin and fabric).

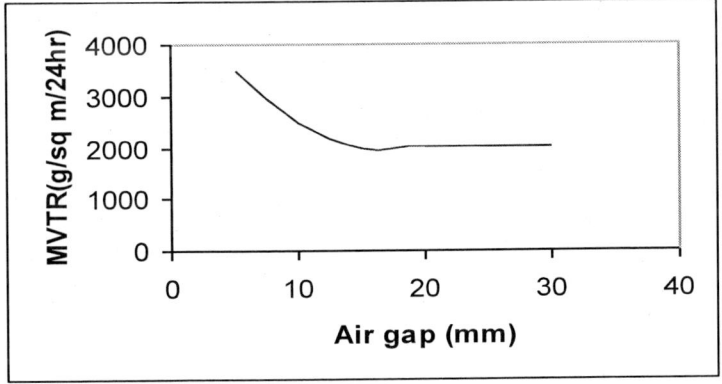

Figure 5.7. Effect of the internal air gap size on the moisture Vapor transmission (MVTR).

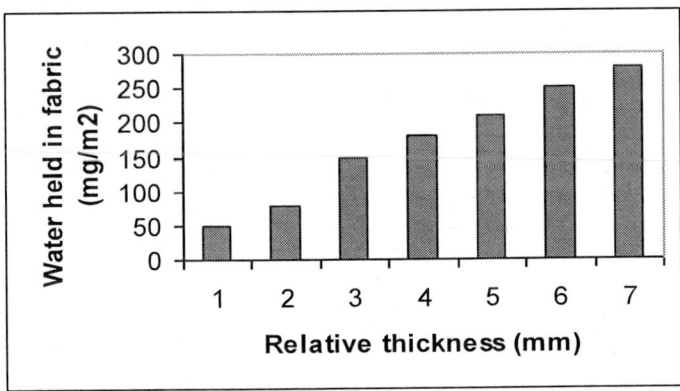

Figure 5.8. Thickness versus the amount of water held in fabric.

REFERENCES

[1] A.R. Horrocks and S.C. Anand, *Handbook of technical textiles*, Woodhead publishing Ltd, England, 2000.

[2] Hollies, N.R.S., Mt. *Washington feasibility test*, Report #20 contract DA-19-129-qm 331, Natick R&D Center, June 1956, AD 698, 450.

[3] Hollies, N.R.S., *Cotton clothing attributes in subject comfort, 15th textile chemistry and processing conference,* USDA New Orleans, 1975.

[4] Hollies, N.R.S., Psycological scaling in comfort assessment, ch. 8 in "Clothing Comfort", N.R.S. Hollies and R.F. Goldman, Eds, *Ann. Arbor Science, Ann. Arbor,* 1977.

[5] D.M. Scheurell, S.M. Spivak, and N. R.S. Hollies, Dynamic surface wetness of fabrics in relation to clothing comfort, *Text Res. J.*, 394-399, 1985.

[6] D.A.De Vries, *Transactions, American Geophysical Union,* 39(5), 909-916, 1958.

[7] J.R. Philip and D.A. DE Vries, *Transactions, American Geophysical Union,* 38, 222-232, 1957.

[8] Eckert E. and Faghri M. (1980): A general analysis of moisture migration caused by temperature difference in an unsaturated porous medium, Int. J. Heat Mass Transfer, vol. 23, pp.1613-1623. *Int. J. Heat Mass Transfer,* 23, 1613-1623, 1980.

[9] Udell K.S. (1985): Heat transfer in porous media considering phase change and capillarity-the heat pipe effect, Int. *J. Heat Mass Transfer,* vol. 28(2), pp.485-495.

[10] Udell K.S. (1983): Heat transfer in porous media heat from above with evaporation, condensation and capillary effects, *J. Heat Transfer,* vol.105, pp.485-492.

[11] Bouddour A. , Auriault J.L. , Mhamadi M. and Bloch J.F. (1998): Heat and mass transfer in wet porous media in presence of evaporation-condensation, *Int. J. Heat Mass Transfer,* vol. 41(15), pp. 2263-2277.

[12] P.W. Gibson, In *Computational Technologies for Fluid/Thermal/ Structural/Chemical Systems with Industrial Applications,* Volume II, pages 125-139, ASME, 1999.

[13] Nordon P. and David H.G. (1967): Coupled diffusion of moisture and heat in hygroscopic textile materials, *Int. J. Heat Mass Transfer*, vol. 10, pp. 853-866.

[14] Farnworth B. (1983): Mechanics of heat flow through clothing insulation, *Tex. Res. J.* vol. 56, pp. 581-587.

[15] Farnworth B. (1986): A numerical model of the combined diffusion of heat and water vapor through clothing. - *Tex. Res. J.,* vol. 56, pp.653-665.

[16] Osczevski, R.J. and Dolhan, P.A., J. Coated Fabrics 18, 255-258 (1989). Osczevski, R.J. (1966): Water Vapour Transfer Through a Hydrophilic Film at Subzero Temperatures, *Tex. Res. J.* vol. 66(1), pp. 24-29.

[17] Farnworth B., Lotens, W.A., and Wittgen, P. (1990): Variation of Water Vapour Resistance of Microporous and Hydrophilic Films with Relative Humidity, *Tex. Res. J. vol.* 60(1), pp. 50-53.

[18] Osczevski R.J. and Dolhan, P.A.(1989): Anomatous Diffusion in a Water Vapor Permeable, waterproff Coating, *J. Coated Fabrics* vol. 18, pp. 255-258.

[19] Gretton, J.C., Brook, D.B., Dyson, H.M., and Harlock, S.C. (1998): Moisture Vapour transport Through Waterproof Breathable Fabrics and Clothing Systems Under a Temperature Gradient, *Tex. Res, J.* vol. 68(12), pp. 936-941.

[20] Galbraith R.L., Werden, J.E., Fahnestock, M.K. (1962): Comfort of Subjects Clothed in Cotton, Water Repellent Cotton and Orlon Suits, *Tex. Res. J.,* vol. 32, pp. 236-243.

[21] Niwa M. (1968): Water Vapor Permeability of Underwear, J. Jpn. Res. Assn. *Textile End Uses.* Vol. 9, pp.446 450.

[22] Morooka, H., and Niwa, M., Moisture and Water Transport Properties of Clothing Materials and Comfort Sensations, *J. Home Econ. Jpn,* 30, 320 (1979)

[23] Hollies, N., Improved Comfort Polyester, Textile Res. J. 54, 544(1984).

[24] King G. and Cassie A.,(1940): Propagation of temperature changes through textiles in humid atmospheres. *Trans Faraday Soc.,* vol. 36, pp. 445-453.

[25] Gretton J.C. and Brook D.B. (2000): Moisture vapor Transport through Clothing System, *Tex. Res. J.,* vol.56, pp.872-879.

[26] J. Crank, The Mathematics of Diffusion, Clarendon Press, Oxford, UK 1975.

[27] P. Henry, Proc. Roy. Soc., 171A(1939) 215., Diffusion in absorbing media.

[28] P. Henry, Disc. Faraday Soc., 3(1948) 243., The diffusion of moisture and heat through textiles.

[29] Nielsen, R., and Edrusick, T.L. (1990): Thermoregulatory Responses to Intermittent Exercise are Influenced by Knit Structure of Underwear, *Eur. J. Appl. Physiol.,* vol.60, pp.15-25.

[30] J. Wehner, B. Miller and L. Rebenfeld, Dynamics of water vapour transmission through fabric barrier, *Tex. Res. J.,* 58(1988) 581.

[31] Y. Li, and B. Holcombe, A two stage sorption model of the coupled diffusion of moisture and heat in wool fabrics, *Text. Res. J.,* 62 (1992) 211.

[32] Y. Li and Z. Luo, An improved mathematical simulation of the coupled diffusion of moisture and heat in wool fabric, *Text. Res. J.,* 69(1999) 760.

[33] Holmer I. (1985): A study of heat transfer through fabrics, *Tex. Res. J.,* vol. 55,pp.511-518.

[34] Snycerski M and Wasiak I.F. (2002): Influence of Furniture Covering Textiles on Moisture Transport in a Car Seat Upholstery Package, *Autex Res. J.* vol. 2 , pp.126-131.

[35] Bakkerig M.K. and Nielsen R. (1995): Some aspects of clothing comfort.-*Ergonomics vol.* 38, pp.926-934.

[36] Galbraith R.L., Werden, J.E., Fahnestock, M.K. (1962): Comfort of Subjects Clothed in Cotton, Water Repellent Cotton and Orlon Suits, *Tex. Res. J.,* vol. 32, pp. 236-243.

[37] Niwa M. (1968): Water Vapor Permeability of Underwear, *J. Jpn. Res. Assn. Textile End Uses.* Vol. 9, pp.446 450.

[38] Hollies, N., Demartino, R., Yoon, H., Buckley,a., Becker, C., and Jackson, W. (1984): Improved Comfort Polyester, *Tex. Res. J.,* vol. 54, pp.544-551.

[39] M. Adler, and W.K. Walsh, Mechanisms of Moisture Transport between Fabrics, *Tex. Res. J.,*84 (1984), 334-343.

[40] M. Day, and P.Z. Sturgeon, Water Vapor Transmission Rates Throuh Materials as Measured by Differential Scanning Calorimetry, *Tex. Res. J.,* 86 (1986)157-161.

[41] A.M. Schneider, and B.N.Hoschke, Heat Transfer Through Moist Fabrics, *Textile Res. J.,* 62 (1992), 61-66.

[42] P. Gibson, D. Rivin, C. Kendrick, and H. Schreuder, Humidity-Dependent Air Permeability of Textile Materials, *Textile Res. J.,* 69(1999), 311-317.

[43] R.M.Crow and R. J. Osczevski, The Interaction of Water with Fabrics, *Textile Res. J.,* 68(1998), 280-288.

[44] M. Snycerski, I.F. Wasiak, Influence of Furniture Covering Textiles on Moisture Transport in a Car Seat Upholstery Package, *Autex Res. J.* 2(2002), 126-131.

[45] K.Prasad, W.Twilley, and J.R. Lawson, Thermal Performance of Fire Fighters' Protective Clothing, NISTIR Report No 6881, 1-32, (2004).

[46] M. Sozen and K. Vafai, Analysis of the non-thermal equilibrium condensing Flow of a Gas Through a Packed bed, *International J. of Heat Mass Transfer,* 33, 1990, 1247-1261.

[47] J.J.Barry, R.W. Hill, Computational Modeling of Protective Clothing, *INJ Report,* 2003, 25-34.
[48] S. Whitaker, in advances in Heat Transfer, Vol.31, edited by J. Hartnett, (Academic Press, New York, 1998), p.1.
[49] Holcombe, B.V., and Hoschke, B.N., Dry Heat transfer characteristics of underwear fabrics, *Textile Res. J.* 53, 368-374 (1983).
[50] Marsh, M.C., The thermal insulating properties of fabrics, *proc. Phys. Soc.* (London) 42, 570 (1930).
[51] Monego, C.J., Golub, S. J., Insulating values of fabrics, Foams and Laminates, *Am. Dyest. Rep.* 52(1), 21-32 (1963).
[52] Morris, M.A., Thermal Insalation of single and multiple layer of fabrics, *Textile Res. J.* 25, 766-773 (1955).
[53] Peirce, F. T., and Rees, W.H., The transmission of heat through textile fabrics, Part II, *J. Textile Inst.* 37, T181-T204 (1946).,
[54] Rees, W.H., The protective value of Clothing, *J. Textile Inst,* 37, 132-152 (1946).
[55] T. Yasuda, M. Miyama and H. Yasuda, Dynamic water vapour and heat transport through layered fabrics, *Textile Res. J.,* 62(4), 227-235, 1992.
[56] Fourt, L., Stooknee, A.M., Fisherman, D., and Harris, M., The rate of drying of fabrics, *Textile Res. J.* 21, 26-33 (1951).
[57] Coplan, M. J., Some moisture relations of Wool and several synthetic fibers and blends, *Textile Res. J.* 23, 897-916 (1953).
[58] Crow, R. M., and Dewar, M. M., Wicking, Regain, Hydrophilicity and drying of textiles, Defence Research Establishment Ottawa Report No. 1180, 1993.
[59] Crow, R.M. and Dewar, M.M., The effect of fiber and fabric properties on fabric drying times, Defense Research Establishment Ottawa Report No. 1182, 1993.
[60] Crow, R. M., and Dewar, M.M., Liquid transport across fabric layers, Defense Research Establishment Ottawa Report No. 1002, 1989.
[61] Holmer, I, Heat exchange and thermal insulation compared in woolen and nylon garments during wearing trial, *Textile Res. J.* 55, 511-518, 1985.
[62] A.H. Woodcock, Moisture transfer in textile systems, *Textile Res. J.* 32, 719-723, 1962.
[63] Henry, P.S.H., The diffusion in absorbing media, *Proc. Roy. Soc.,* 171, 215-241, 1939.
[64] Henry, P.S.H, The diffusion of moisture and heat through textiles, *Discuss. Farad. Soc.* 3, 243-257, 1948.

[65] David, H.G., and Nordon, P., Case studies of coupled heat and moisture diffusion in wool beds, *Textile Res. J.* 39, 166-172 (1969).

[66] Farnworth, B., A numerical model of the combined diffusion of heat and water vapour through clothing, *Textile Res. J.* 56, 653-665, 1986.

[67] Wehner, J. A., Miller, B., and Rebenfeld, L., Dynamics of water vapour transmission through fabric barriers, *Textile Res. J.* 58, 581-592, 1988.

[68] Downes, J. G., and Mackay, B. H., Sorption kinetics of water vapour in wool fibers, *J. Polym. Sci.*, 28, 45-67, 1958.

[69] Watt, I.C., Kinetic study of the wool-water system, Part I: the mechanisms of two-stage sorption, *Textile Res. J.* 58, 581-592, 1988.

APPENDIX

HEAT TRANSFER

Heat transfer is the study of energy movement in the form of heat which occurs in many types of processes. The transfer occurs from the high to the low temperature regions. Therefore a temperature gradient has to exist between the two regions for heat transfer to happen. It can be done by conduction (within one solid or between two solids in contact), by convection (between two fluids or a fluid and a solid in direct contact with the fluid), by radiation (transmission by electromagnetic waves through space) or by combination of the above three methods.

The general equation for heat transfer is:

$$\begin{pmatrix} \text{rate of} \\ \text{heat in} \end{pmatrix} + \begin{pmatrix} \text{rate of generation} \\ \text{of heat} \end{pmatrix} = \begin{pmatrix} \text{rate of} \\ \text{heat out} \end{pmatrix} + \begin{pmatrix} \text{rate of accumulation} \\ \text{of heat} \end{pmatrix}$$

MASS TRANSFER

Textiles exposed to a hot air stream may be cooled evaporatively by bleeding water through its surface. Water vapour may condense out of damp air onto cool surfaces. Heat will flow through an air-water mixture in these situations, but water vapour will diffuse through air as well. This sort of transport of one substance relative to another called mass transfer.

DRYING OF TEXTILES

Drying of textiles is accomplished by vaporizing the water and to do this the latent heat of vaporization must be supplied. There are, thus, two important process-controlling factors that enter into the process of drying:

(a) transfer of heat to provide the necessary latent heat of vaporization,
(b) movement of water or water vapor through textiles and then away from it to effect separation of water.

Drying processes fall into different categories:
In air and contact drying, heat is transferred through the textiles either from heated air or from heated surfaces.
The water vapor is removed with the air. Heat transfer in is generally by convection, conduction, sometimes by radiation.

STEADY STATE HEAT TRANSFER

Heat transfer is said to be at steady-state when the quantity of heat flowing from one point to another by unit time is constant and the temperatures at each point in the system do not change with time.
Assuming no heat generation, no accumulation of heat and transfer of heat by conduction, at steady state we have:

$$q_x = q_{x+\Delta x}$$

$$q_x = -kA\frac{dT}{dx}$$

q_x/A is the heat flux in Wm^{-2} while the quantity dT/dx represent the temperature gradient in the x direction.

Appendix

CONVECTIVE HEAT-TRANSFER COEFFICIENT

When a fluid is in forced or natural convective motion along a surface, the rate of heat transfer between the solid and the fluid is expressed by the following equation:

$$q = h.A(T_W - T_f)$$

The coefficient h is dependent on the system geometry, the fluid properties and velocity and the temperature gradient. Most of the resistance to heat transfer happens in the stationary layer of fluid present at the surface of the solid, therefore the coefficient h is often called film coefficient.

DIMENSIONLESS NUMBERS IN CONVECTIVE HEAT TRANSFER

Correlations for predicting film coefficient h are semi empirical and use dimensionless numbers which describe the physical properties of the fluid, the type of flow, the temperature difference and the geometry of the system.

The Reynolds Number characterizes the flow properties (laminar or turbulent). L is the characteristic length: length for a plate, diameter for cylinder or sphere.

$$N_{Re} = \frac{\rho L v}{\mu}$$

The Prandtl Number characterizes the physical properties of the fluid for the viscous layer near the wall.

$$N_{Pr} = \frac{\mu c_p}{k}$$

The Nusselt Number relates the heat transfer coefficient h to the thermal conductivity k of the fluid.

$$N_{Nu} = \frac{hL}{k}$$

The Grashof Number characterizes the physical properties of the fluid for natural convection.

$$N_{Gr} = \frac{L^3 \Delta \rho g}{\rho \gamma^2} = \frac{L^3 \rho^2 g \beta \Delta T}{\mu^2}$$

RADIATION HEAT TRANSFER

Radiation is a term applied to many processes which involve energy transfer by electromagnetic wave (x rays, light, gamma rays ...). It obeys the same laws as light, travels in straight lines and can be transmitted through space and vacuum. It is an important mode of heat transfer encountered where large temperature difference occurs between two surfaces such as in furnaces, radiant driers and baking ovens.

The thermal energy of the hot source is converted into the energy of electromagnetic waves. These waves travel through space into straight lines and strike a cold surface. The waves that strike the cold body are absorbed by that body and converted back to thermal energy or heat. When thermal radiations falls upon a body, part is absorbed by the body in the form of heat, part is reflected back into space and in some case part can be transmitted through the body.

The basic equation for heat transfer by radiation from a body at temperature T is:

$$q = A\varepsilon\sigma T^4$$

where ϵ is the emissivity of the body. $\epsilon = 1$ for a perfect black body while real bodies which are gray bodies have an $\epsilon < 1$

Combining Radiation and Convection Heat Transfer

In many applications radiation and convection occurs at the same time. A body receiving energy from radiation will return some to its surrounding (unless in vacuum) through convective heat transfer.

For convenience, a radiation heat transfer coefficient h_R expressed in W/m^2.K can be evaluated in the following manner:

$$q = A \varepsilon \sigma \left(T_1^4 - T_2^4 \right) = h_r A \left(T_1 - T_2 \right)$$

$$h_r = \varepsilon (5.67 \times 10^{-8}) \frac{\left(T_1^4 - t_2^4 \right)}{\left(T_1 - T_2 \right)}$$

Conductive Heat Transfer

Fourier's Law can be integrated through a flat wall of constant cross section A for the case of steady-state heat transfer when the thermal conductivity of the wall k is constant.

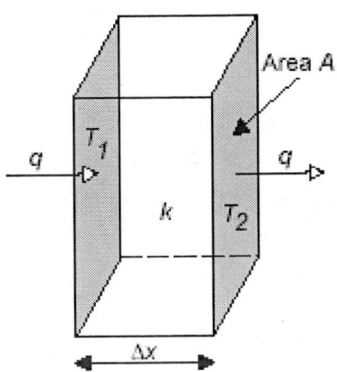

$$\frac{q}{A} \int_{x_1}^{x_2} dx = -k \int_{T_1}^{T_2} dT \rightarrow \frac{q}{A} = \frac{k}{\Delta x} (T_1 - T_2)$$

At any position x between x_1 and x_2, the temperature T varies linearly withthe distance:

$$\frac{q}{A} = \frac{k}{x - x_1}(T_1 - T)$$

GENERAL CONCEPTS

For Heat flow analysis of wet porous nanostructure fabrics, the liquid is water and the gas is air. Evaporation or condensation occurs at the interface between the water and air so that the air is mixed with water vapor. A flow of the mixture of air and vapor may be caused by external forces, for instance, by an imposed pressure difference. The vapor will also move relative to the gas by diffusion from regions where the partial pressure of the vapor is higher to those where it is lower.

Heat flow in porous nanostructure fabrics is the study of energy movement in the form of heat which occurs in many types of processes. The transfer of heat in porous nanostructure fabrics occurs from the high to the low temperature regions. Therefore a temperature gradient has to exist between the two regions for heat transfer to happen. It can be done by conduction (within one porous solid or between two porous solids in contact), by convection (between two fluids or a fluid and a porous solid in direct contact with the fluid), by radiation (transmission by electromagnetic waves through space) or by combination of the above three methods.

The general equation for heat transfer in porous media is:

$$\begin{pmatrix}\text{rate of}\\ \text{heat in}\end{pmatrix} + \begin{pmatrix}\text{rate of generation}\\ \text{of heat}\end{pmatrix} = \begin{pmatrix}\text{rate of}\\ \text{heat out}\end{pmatrix} + \begin{pmatrix}\text{rate of accumulation}\\ \text{of heat}\end{pmatrix}$$

When a wet porous nanostructure fabrics material is subjected to thermal drying two processes occur simultaneously, namely:

a) Transfer of heat to raise the wet porous media temperature and to evaporate the moisture content.
b) Transfer of mass in the form of internal moisture to the surface of the porous material and its subsequent evaporation.

The rate at which drying is accomplished is governed by the rate at which these two processes proceed. Heat is a form of energy that can across the boundary of a system. Heat can, therefore, be defined as "the form of energy that is transferred between a system and its surroundings as a result of a temperature difference". There can only be a transfer of energy across the boundary in the form of heat if there is a temperature difference between the system and its surroundings. Conversely, if the system and surroundings are at the same temperature there is no heat transfer across the boundary.

Strictly speaking, the term *"heat"* is a name given to the particular form of energy crossing the boundary. However, heat is more usually referred to in thermodynamics through the term "heat transfer", which is consistent with the ability of heat to raise or lower the energy within a system.

There are three modes of heat flow in porous nanostructure fabrics media:

- Convection
- Conduction
- Radiation

All three are different. Convection relies on movement of a fluid in porous material. Conduction relies on transfer of energy between molecules within a porous solid or fluid. Radiation is a form of electromagnetic energy transmission and is independent of any substance between the emitter and receiver of such energy. However, all three modes of heat flow rely on a temperature difference for the transfer of energy to take place.

The greater the temperature difference the more rapidly will the heat be transferred. Conversely, the lower the temperature difference, the slower will be the rate at which heat is transferred. When discussing the modes of heat transfer it is the rate of heat transfer Q that defines the characteristics rather than the quantity of heat.

As it was mentioned earlier, there are three modes of heat flow in porous structures, convection, conduction and radiation. Although two, or even all three, modes of heat flow may be combined in any particular thermodynamic situation, the three are quite different and will be introduced separately.

The coupled heat and liquid moisture transport of porous material has wide industrial applications. Heat transfer mechanisms in porous textiles include conduction by the solid material of fibers, conduction by intervening air, radiation, and convection. Meanwhile, liquid and moisture transfer mechanisms include vapor diffusion in the void space and moisture sorption by the fiber, evaporation, and capillary effects. Water vapor moves through porous textiles as a result of water vapor concentration differences. Fibers absorb water vapor due to their internal chemical compositions and structures. The flow of liquid moisture through the textiles is caused by fiber-liquid molecular attraction at the surface of fiber materials, which is determined mainly by surface tension and effective capillary pore distribution and pathways. Evaporation and/or condensation take place, depending on the temperature and moisture distributions. The heat transfer process is coupled with the moisture transfer processes with phase changes such as moisture sorption/desorption and evaporation/condensation.

DRYING OF POROUS NANOSTRUCTURE FABRICS

All three of the mechanisms by which heat is transferred- conduction, radiation and convection, may enter into drying. The relative importance of the mechanisms varies from one drying process to another and very often one mode of heat transfer predominates to such extent that it governs the overall process.

As an example, in air drying the rate of heat transfer is given by:

$$q = h_s A(T_a - T_s) \qquad (A.1)$$

Where q is the heat transfer rate in Js^{-1}, h_s is the surface heat-transfer coefficient in $Jm^{-2} s^{-1} \,°C^{-1}$, A is the area through which heat flow is taking place, m^{-2}, T_a is the air temperature and T_s is the temperature of the surface which is drying, °C.

To take another example, in a cylindrical dryer where moist material is spread over the surface of a heated cylinder, heat transfer occurs by conduction from the cylinder to the porous media, so that the equation is

$$q = UA(T_i - T_s) \qquad (A.2)$$

Where U is the overall heat-transfer coefficient, T_i is the cylinder temperature (usually very close to that of the steam), Ts is the surface temperature of textile and A is the area of the drying surface on the cylinder. The value of U can be estimated from the conductivity of the cylinder material and of the layer of porous solid.

Mass transfer in the drying of a wet porous material will depend on two mechanisms: movement of moisture within the porous material which will be a function of the internal physical nature of the solid and its moisture content; and the movement of water vapor from the material surface as a result of water vapour from the material surface as a result of external conditions of temperature, air humidity and flow, area of exposed surface and supernatant pressure.

Some porous materials such as textiles exposed to a hot air stream may be cooled evaporatively by bleeding water through its surface. Water vapor may condense out of damp air onto cool surfaces. Heat will flow through an air-water mixture in these situations, but water vapor will diffuse through air as well. This sort of transport of one substance relative to another called mass transfer. The moisture content, X, is described as the ratio of the amount of water in the materials, m_{H2O} to the dry weight of material, $m_{material}$:

$$X = \frac{m_{H2O}}{m_{material}} \qquad (A\text{-}3)$$

There are large differences in quality between different porous materials depending on structure and type of material. A porous material such as textiles can be hydrophilic or hydrophobic. The hydrophilic fibres can absorb water, while hydrophobic fibers do not. A textile that transports water through its porous structures without absorbing moisture is preferable to use as a first layer. Mass transfer during drying depends on the transport within the fiber and from the textile surface, as well as on how the textile absorbs water, all of which will affect the drying process.

As the critical moisture content or the falling drying rate period is reached, the drying rate is less affected by external factors such as air velocity. Instead, the internal factors due to moisture transport in the material will have a larger impact. Moisture is transported in textile during drying through

- Capillary flow of unbound water
- Movement of bound water and
- Vapor transfer

Unbound water in a porous media such as textile will be transported primarily by capillary flow.

As water is transported out of the porous material, air will be replacing the water in the pores. This will leave isolated areas of moisture where the capillary flow continues.

Moisture in a porous structure can be transferred in liquid and gaseous phases. Several modes of moisture transport can be distinguished:

- Transport by liquid diffusion
- Transport by vapor diffusion
- Transport by effusion (Knudsen-type diffusion)
- Transport by thermo-diffusion
- Transport by capillary forces
- Transport by osmotic pressure and
- Transport due to pressure gradient.

CONVECTION HEAT FLOW IN POROUS MEDIA

A very common method of removing water from porous structures is convective drying. Convection is a mode of heat transfer that takes place as a result of motion within a fluid. If the fluid, starts at a constant temperature and the surface is suddenly increased in temperature to above that of the fluid, there will be convective heat transfer from the surface to the fluid as a result of the temperature difference. Under these conditions the temperature difference causing the heat transfer can be defined as:

ΔT = surface temperature-mean fluid temperature

Using this definition of the temperature difference, the rate of heat transfer due to convection can be evaluated using Newton's law of cooling:

$$Q = h_c A \Delta T \qquad (A\text{-}4)$$

where A is the heat transfer surface area and h_c is the coefficient of heat transfer from the surface to the fluid, referred to as the "convective heat transfer coefficient".

The units of the convective heat transfer coefficient can be determined from the units of other variables:

$$Q = h_c A \Delta T$$
$$W = (h_c) m^2 K \quad \text{(A-5)}$$

so the units of h_c are $W/m^2 K$.

The relationships given in equations (0.4 and 0.5) are also true for the situation where a surface is being heated due to the fluid having higher temperature than the surface. However, in this case the direction of heat transfer is from the fluid to the surface and the temperature difference will now be

ΔT = mean fluid temperature-surface temperature

The relative temperatures of the surface and fluid determine the direction of heat transfer and the rate at which heat transfer take place.

As given in previous equations, the rate of heat transfer is not only determined by the temperature difference but also by the convective heat transfer coefficient h_c. This is not a constant but varies quite widely depending on the properties of the fluid and the behavior of the flow. The value of h_c must depend on the thermal capacity of the fluid particle considered, i.e. mC_p for the particle.

Two common heat transfer fluids are air and water, due to their widespread availability. Water is approximately 800 times more dense than air and also has a higher value of C_p. If the argument given above is valid then water has a higher thermal capacity than air and should have a better convective heat transfer performance. This is borne out in practice because typical values of convective heat transfer coefficients are as follows:

Fluid	$h_c (W/m^2 K)$
water	500-10000
air	5-100

The variation in the values reflects the variation in the behavior of the flow, particularly the flow velocity, with the higher values of h_c resulting from higher flow velocities over the surface.

When a fluid is in forced or natural convective motion along a surface, the rate of heat transfer between the solid and the fluid is expressed by the following equation:

$$q = h.A(T_W - T_f) \tag{A-6}$$

The coefficient h is dependent on the system geometry, the fluid properties and velocity and the temperature gradient. Most of the resistance to heat transfer happens in the stationary layer of fluid present at the surface of the solid, therefore the coefficient h is often called film coefficient.

Correlations for predicting film coefficient h are semi empirical and use dimensionless numbers which describe the physical properties of the fluid, the type of flow, the temperature difference and the geometry of the system.

The Reynolds Number characterizes the flow properties (laminar or turbulent). L is the characteristic length: length for a plate, diameter for cylinder or sphere.

$$N_{Re} = \frac{\rho L v}{\mu} \tag{A-7}$$

The Prandtl Number characterizes the physical properties of the fluid for the viscous layer near the wall.

$$N_{Pr} = \frac{\mu c_p}{k} \tag{A-8}$$

The Nusselt Number relates the heat transfer coefficient h to the thermal conductivity k of the fluid.

$$N_{Nu} = \frac{hL}{k} \tag{A-9}$$

The Grashof Number characterizes the physical properties of the fluid for natural convection.

$$N_{Gr} = \frac{L^3 \Delta \rho g}{\rho \gamma^2} = \frac{L^3 \rho^2 g \beta \Delta T}{\mu^2} \qquad (A-10)$$

CONDUCTION HEAT FLOW IN POROUS MATERIALS

If a fluid could be kept stationary there would be no convection taking place. However, it would still be possible to transfer heat by means of conduction. Conduction depends on the transfer of energy from one molecule to another within the heat transfer medium and, in this sense, thermal conduction is analogous to electrical conduction.

Conduction can occur within both porous solids and fluids. The rate of heat transfer depends on a physical property of the particular porous solid of fluid, termed its thermal conductivity k, and the temperature gradient across the porous medium. The thermal conductivity is defined as the measure of the rate of heat transfer across a unit width of porous material, for a unit cross-sectional area and for a unit difference in temperature.

From the definition of thermal conductivity k it can be shown that the rate of heat transfer is given by the relationship:

$$Q = \frac{kA\Delta T}{x} \qquad (A-11)$$

where ΔT is the temperature difference $T_1 - T_2$, defined by the temperature on the either side of the porous solid. The units of thermal conductivity can be determined from the units of the other variables:

$$\begin{aligned} Q &= kA\Delta T / x \\ W &= (k)m^2 K / m \end{aligned} \qquad (A-12)$$

so the unit of k are $W/m^2 K/m$, expressed as W/mK.

Fourier's Law can be integrated through a flat wall of constant cross section A for the case of steady-state heat transfer when the thermal conductivity of the wall k is constant.

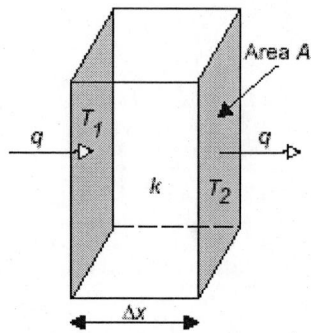

$$\frac{q}{A}\int_{x_1}^{x_2}dx = -k\int_{T_1}^{T_2}dT \rightarrow \frac{q}{A} = \frac{k}{\Delta x}(T_1 - T_2) \qquad (A\text{-}13)$$

At any position x between x_1 and x_2, the temperature T varies linearly with the distance:

$$\frac{q}{A} = \frac{k}{x - x_1}(T_1 - T) \qquad (A\text{-}14)$$

RADIATION HEAT FLOW IN POROUS SOLIDS

The third mode of heat flow, radiation, does not depend on any medium for its transmission. In fact, it takes place most freely when there is a perfect vacuum between the emitter and the receiver of such energy. This is proved daily by the transfer of energy from the sun to the earth across the intervening space.

Radiation is a form of electromagnetic energy transmission and takes place between all matters providing that it is at a temperature above absolute zero. Infra-red radiation form just part of the overall electromagnetic spectrum. Radiation is energy emitted by the electrons vibrating in the molecules at the surface of a porous body. The amount of energy that can be transferred depends on the absolute temperature of the porous body and the radiant properties of the surface.

A porous body that has a surface that will absorb all the radiant energy it receives is an ideal radiator, termed a "black body". Such a porous body will not only absorb radiation at a maximum level but will also emit radiation at a maximum level. However, in practice, porous bodies do not have the surface characteristics of a black body and will always absorb, or emit, radiant energy at a lower level than a black body.

It is possible to define how much of the radiant energy will be absorbed, or emitted, by a particular surface by the use of a correction factor, known as the "emissivity" and given the symbol ε. The emmisivity of a surface is the measure of the actual amount of radiant energy that can be absorbed, compared to a black body. Similarly, the emissivity defines the radiant energy emitted from a surface compared to a black body. A black body would, therefore, by definition, have an emissivity ε of 1.

Since World War II, there have been major developments in the use of microwaves for heating applications. After this time it was realized that microwaves had the potential to provide rapid, energy-efficient heating of materials. These main applications of microwave heating today include food processing, wood drying, plastic and rubber treating as well as curing and preheating of ceramics. Broadly speaking, microwave radiation is the term associated with any electromagnetic radiation in the microwave frequency range of 300 MHz-300 Ghz. Domestic and industrial microwave ovens generally operate at a frequency of 2.45 Ghz corresponding to a wavelength of 12.2 cm. However, not all materials can be heated rapidly by microwaves. Porous materials may be classified into three groups, *i.e.* conductors insulators and absorbers. Porous materials that absorb microwave radiation are called dielectrics, thus, microwave heating is also referred to as dielectric heating. Dielectrics have two important properties:

- They have very few charge carriers. When an external electric field is applied there is very little change carried through the material matrix.
- The molecules or atoms comprising the dielectric exhibit a dipole movement distance.

An example of this is the stereochemistry of covalent bonds in a water molecule, giving the water molecule a dipole movement. Water is the typical case of non-symmetric molecule. Dipoles may be a natural feature of the dielectric or they may be induced. Distortion of the electron cloud around non-polar molecules or atoms through the presence of an external electric field can induce a temporary dipole movement. This movement generates friction inside the dielectric and the energy is dissipated subsequently as heat.

The interaction of dielectric materials with electromagnetic radiation in the microwave range results in energy absorbance. The ability of a material to absorb energy while in a microwave cavity is related to the loss tangent of the material.

This depends on the relaxation times of the molecules in the material, which, in turn, depends on the nature of the functional groups and the volume of the molecule. Generally, the dielectric properties of a material are related to temperature, moisture content, density and material geometry.

An important characteristic of microwave heating is the phenomenon of "hot spot" formation, whereby regions of very high temperature form due to non-uniform heating. This thermal instability arises because of the non-linear dependence of the electromagnetic and thermal properties of material on temperature. The formation of standing waves within the microwave cavity results in some regions being exposed to higher energy than others.

Cavity design is an important factor in the control, or the utilization of this "hot spots" phenomenon.

Microwave energy is extremely efficient in the selective heating of materials as no energy is wasted in "bulk heating" the sample. This is a clear advantage that microwave heating has over conventional methods. Microwave heating processes are currently undergoing investigation for application in a number of fields where the advantages of microwave energy may lead to significant savings in energy consumption, process time and environmental remediation.

Compared with conventional heating techniques, microwave heating has the following additional advantages:

- higher heating rates;
- no direct contact between the heating source and the heated material;
- selective heating may be achieved;
- greater control of the heating or drying process;
- reduced equipment size and waste.

As mentioned earlier, radiation is a term applied to many processes which involve energy transfer by electromagnetic wave (x rays, light, gamma rays ...). It obeys the same laws as light, travels in straight lines and can be transmitted through space and vacuum. It is an important mode of heat transfer encountered where large temperature difference occurs between two surfaces such as in furnaces, radiant driers and baking ovens.

The thermal energy of the hot source is converted into the energy of electromagnetic waves. These waves travel through space into straight lines and strike a cold surface. The waves that strike the cold body are absorbed by that body and converted back to thermal energy or heat. When thermal radiations falls upon a body, part is absorbed by the body in the form of heat, part is reflected back into space and in some case part can be transmitted through the body.

The basic equation for heat transfer by radiation from a body at temperature T is:

$$q = A\varepsilon\sigma T^4 \qquad (A\text{-}15)$$

where ϵ is the emissivity of the body. $\epsilon = 1$ for a perfect black body while real bodies which are gray bodies have an $\epsilon < 1$

POROSITY AND PORE SIZE DISTRIBUTION IN A BODY

Porosity refers to volume fraction of void spaces. This void space can be actual space filled with air or space filled with both water and air. Many different definitions of porosity are possible. For non-hygroscopic materials, porosity does not change with change in moisture content. For hygroscopic materials, porosity changes with moisture content. However, such changes during processing are complex due to consideration of bound water and are typically not included in computations.

The distinction between porous and capillary-porous is based on the presence and size of the pores. Porous materials are sometimes defined as those having pore diameter greater than or equal to 10^{-7} m and capillary-porous as one having diameter less than 10^{-7} m. Porous and capillary porous materials were defined as those having a clearly recognizable pore space.

In non-hygroscopic materials, the pore space is filled with liquid if the material is completely saturated and with air if it is completely dry. The amount of physically bound water is negligible. Such a material does not shrink during heating. In non-hygroscopic materials, vapour pressure is a function of temperature only. Examples of non-hygroscopic capillary-porous materials are sand, polymer particles and some ceramics. Transport materials in non-hygroscopic materials do not cause any additional complications as in hygroscopic materials.

In hygroscopic materials, there is large amount of physically bound water and the material often shrinks during heating. In hygroscopic materials there is a level of moisture saturation below which the internal vapour pressure is a function of saturation and temperature. These relationships are called equilibrium moisture isotherms. Above this moisture saturation, the vapour pressure is a function of temperature only and independent of the moisture level. Thus, above certain moisture level, all materials behave non-hygroscopic.

Transport of water in hygroscopic materials can be complex. The unbound water can be in funicular and pendular states. This bound water is removed by progressive vaporization below the surface of the solid, which is accompanied by diffusion of water vapour through the solid.

Examples of porous materials are to be found in everyday life. Soil, porous or fissured rocks, ceramics, fibrous aggregates, sand filters, snow layers and a piece of sugar or bread are but just a few. All of these materials have properties in common that intuitively lead us to classify them into a single denomination: *porous media*.

Indeed, one recognizes a common feature to all these examples. All are described as "solids" with "holes", i.e. presenting *connected void spaces*, distributed - randomly or quite homogeneously - within a *solid matrix*. Fluid flows can occur within the porous medium, so that we add one essential feature: this *void space* consists of a complex tridimensional network of *interconnected* small empty volumes called "*pores*", with several continuous paths linking up the porous matrix spatial extension, to enable flow across the sample.

If we consider a porous medium that is not consolidated, it is possible to derive the *particle-size distribution* of the constitutive solid grains. The problem is obvious when dealing with spherical shaped particles, but raises the question of what is meant by particle size in the case of an irregular shaped particle. In both cases, a first intuitive approach is to start with a *sieve analysis*. It consists to sort the constitutive solid particles among various sieves, each one having a calibrated mesh size. The most common type of sieve is a woven cloth of stainless steel or other metal, with wire diameter and tightness of weave controlled to produced roughly rectangular openings of known, uniform size. By shaking adequately the raw granular material, the solid grains are progressively falling through the stacked sieves of decreasing mesh sizes, i.e. a *sieve column*. We finally get separation of the grains as function of their *particle- size* distribution that is also denoted by the porous medium *granulometry*. This method can be implemented for dry granular samples. The *sieve analysis* is a very simple and inexpensive separation method, but the reported granulometry depends very much on the shape of the particles and the duration of the laboratory test, since the sieve will let in theory pass any particle with a smallest cross-section smaller than the nominal mesh opening. For example, one gets very different figure while comparing long thin particles to spherical particles of the same weight.

The definition of a porous medium can be based on the objective of describing flow in porous media. A porous medium is a heterogeneous system consisting of a rigid and stationary solid matrix and fluid filled voids. The solid matrix or phase is always continuous and fully connected. A phase is considered a homogeneous portion of a system, which is separated from other such portions by a definitive boundary, called an interface. The size of the voids or pores is large enough such that the contained fluids can be treated as a continuum. On the other hand, they are small enough that the interface between different fluids is not significantly affected by gravity.

The topology of the solid phase determines if the porous medium is permeable, i.e. if fluid can flow through it, and the geometry determines the resistance to flown and therefore the permeability. The most important influence of the geometry on the permeability is through the interfacial or surface area between the solid phase and the fluid phase. The topology and geometry also determine if a porous medium is isotropic, i.e. all parameters are independent of orientation or anisotropic if the parameters depend on orientation. In multi- phase flow the geometry and surface characteristics of the solid phase determine the fluid distribution in the pores, as does the interaction between the fluids. A porous medium is homogeneous if its average properties are independent of location, and heterogeneous if they depend on location. An example of a porous medium is sand. Sand is an unconsolidated porous medium, and the grains have predominantly point contact. Because of the irregular and angular nature of sand grains, many wedge-like crevices are present. An important quantitative aspect is the surface area of the sand grains exposed to the fluid. It determines the amount of water which can be held by capillary forces against the action of gravity and influences the degree of permeability.

The fluid phase occupying the voids can be heterogeneous in itself, consisting of any number of miscible or immiscible fluids. If a specific fluid phase is connected, continuous flow is possible. If the specific fluid phase is not connected, it can still have bulk movement in ganglia or drops. For single-phase flow the movement of a Newtonian fluid is described. For two-phase immiscible flow, a viscous Newtonian wetting liquid together with a non-viscous gas are described. In practice these would be water and air.

PORE-SIZE DISTRIBUTION IN POROUS STRUCTURE

A detailed description of the complex tri-dimensional network of pores is obviously impossible to derive. For consolidated porous media, the determination of a *pore-size distribution* is nevertheless useful. For those particular media, it is indeed impossible to handle any particle-size distribution analysis.

One approach to define a *pore size* is in the following way: the *pore diameter* δ at a given point within the pore space is the diameter of the largest sphere that contains this point, while still remaining entirely within the pore space. To each point of the *pore space* such a "diameter" can be attached rigorously, and the *pore-size distribution* can be derived by introducing the *pore-size density function* $\theta(\delta)$ defined as the fraction of the total void space that has a pore diameter comprised between δ and $\delta + d\delta$. This distribution is normalized by the relation:

$$\int_0^\infty \theta(\delta)d\delta = 1 \qquad (A-16)$$

A porous structure should be

- A material medium made of heterogeneous or multiphase matter. At least one of the considered phases is not solid. The solid phase is usually called the *solid matrix*. The space within the porous medium domain that is not part of the *solid matrix* is named *void space* or *pore space*. It is filled by gaseous and/or liquid phases.
- The solid phase should be distributed throughout the porous medium to draw a network of pores, whose characteristic size can vary greatly. Some of the pores comprising the void space must enable the flow across the solid matrix, so that they should then be interconnected.
- The interconnected pore space is often denoted as the *effective pore space*, while unconnected pores may be considered from the hydrodynamic point of view as part of the solid matrix, since those

pores are ineffective as far as flow through the porous medium is concerned. They are *dead-end pores* or *blind pores*, that contain stagnant fluid and no flow occurs through them.

A porous material is a set of pores embedded in a matrix of mostly solid material. The pores are the voids in the material itself. Pores can be isolated or interconnected. Furthermore, a pore can contain a fluid or a vapor, but it can also be empty. If the pore is completely filled with the fluid, it will be called saturated and if it is partially filled, it will be called non-saturated. So the porous material is primarily characterized by the content of its voids and not by the properties of the material itself. Figure A.1 gives a sketch of a porous material.

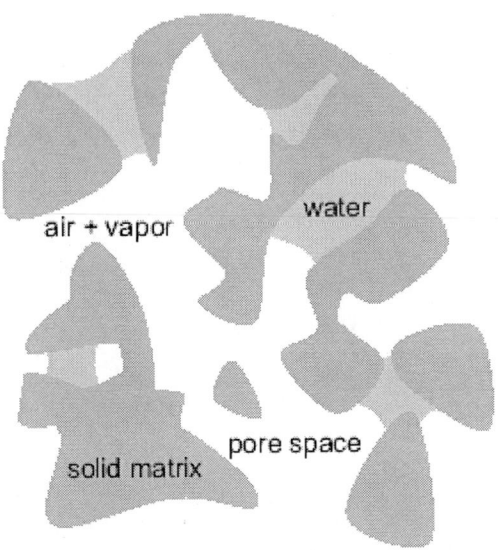

Figure A.1. A 2D sketch of a non-saturated porous material.

If the pores are not interconnected very well, the relaxation-time distribution of an NMR(Nuclear Magnetic Resonance) spin-echo measurement can be interpreted in terms of a pore-size distribution (PSD). For magnetically doped materials like clay and .red-clay this so-called relaxometry technique gives a pore-size distribution between 100 nm and 100 μm, which is also the range of the majority of the pores in these materials. NMR (Nuclear Magnetic Resonance) can be used for spectroscopy, because different nuclei resonate at different frequencies and can therefore be distinguished from each other. Not only nuclei, but also different isotopes can be distinguished. Since also the surrounding of the nucleus has an effect on the exact resonance frequency, NMR spectroscopy is also used to distinguish specific molecules. By manipulating the spatial dependence of the magnetic field strength and the frequency of the RF excitation, the NMR sensitive region can be varied. This enables a noninvasive measurement of the spatial distribution of a certain nucleus and is called NMR Imaging (MRI).

In many NMR experiments it was noticed that liquids confined in porous materials exhibit properties that are very different from those of the bulk fluid. The so-called longitudinal (*T*1) and transverse (*T*2) relaxation time of bulk water, *e.g.*, are on the order of seconds, whereas for water in a porous material these times can be on the order of milliseconds. The measurement of *T*1 and *T*2 in an NMR experiment is often called NMR relaxometry. The transverse relaxation time is more sensitive to local magnetic field gradients inside the porous material than the longitudinal relaxation time. This sensitivity can be used to measure the self-di.usion coe.cient of the liquid. The interpretation of the measured self-diffusion coefficient of a confined liquid is often called NMR diffusometry.

Nuclear Magnetic Resonance is based on the following principle. When a nucleus is placed in a static magnetic field, the nuclear spin $\underline{I}$ will start to presses around this field, since the magnetic moment $\bar{\mu}$ of the nucleus is related to the nuclear spin $\bar{I}$ (Fig. A.2).

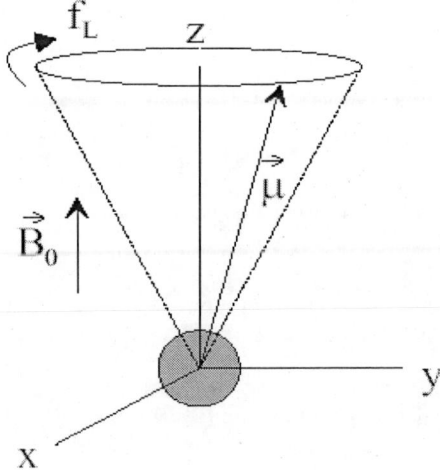

Figure A.2. Larmor precession of a nuclear magnetic moment in a magnetic field.

The frequency of this precession motion is called the Larmor frequency:

$$f_L = \frac{\gamma}{2\pi} B_0 \qquad (A-17)$$

where B_0 is the magnitude of the static magnetic field, which is usually taken aligned with the z-axis, f_L is the Larmor frequency and γ is the gyromagnetic ratio of the nucleus.

The NMR resonance condition (Eq. A-17) states that the Larmor frequency depends linearly on the magnetic field. Normally one starts to assume that the magnetic field in the porous material is equal to the magnetic field generated by the experimental setup. This can be either the magnetic field emerging from a permanent magnet, an electromagnet, or a superconducting magnet. Frequently, an extra magnetic field gradient is added to the main magnetic field. This magnetic field gradient is used to discriminate spins at a certain position from spins at other positions. It is the basic principle of NMR Imaging (MRI). However, the magnetic field inside the porous sample can deviate largely from the magnetic field applied externally.

Because the magnetic susceptibility of the porous material differs from that of the surrounding air, the magnetic field inside the porous sample will deviate from the magnetic field that is present in the sample chamber or insert. Apart from this, the magnetic field in the pores of the material may differ from that in the bulk matrix. Consider two media with a different susceptibility. If the magnetic susceptibility of the sphere is larger (Fig. A.3 on the left) than that of the environment, the magnetic field inside this sphere is larger than the external magnetic field and the sphere is called paramagnetic. If, on the other hand, the susceptibility of the sphere is smaller (Fig. A.3 on the right) than that of the environment, the magnetic field inside the sphere is smaller than the external magnetic field and the sphere is called diamagnetic.

The amount of water in a porous body such as the textiles at the EMC is defined as bound water and it is absorbed by the textile fibers. When the textile is unable to absorb more water, all excess water is defined as unbound moisture. The unbound moisture is often found as a continuous liquid within the porous material.

Drying of porous media is accomplished by vaporizing the water and to do this the latent heat of vaporization must be supplied. There are, thus, two important process-controlling factors that enter into the process of drying:

(a) transfer of heat to provide the necessary latent heat of vaporization, (b) movement of water or water vapor through textiles and then away from it to effect separation of water.

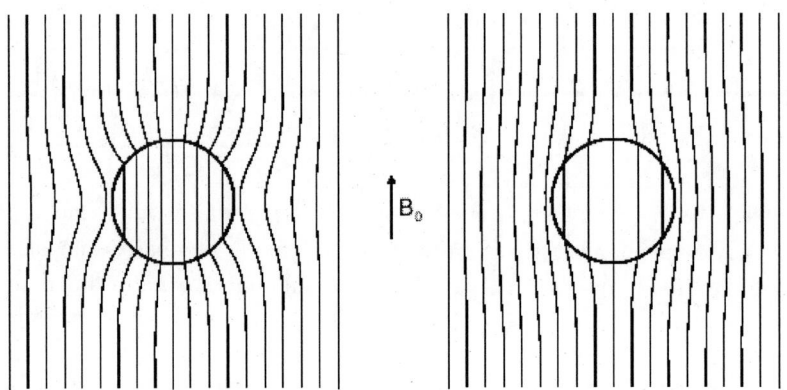

Figure A.3. Disturbance of homogeneous magnetic field B_0 by an object with different susceptibility. Plotted are the magnetic field lines. On the left: a paramagnetic sphere; on the right: a diamagnetic sphere.

BASIC FLOW RELATIONS IN POROUS BODY

The motion of a fluid is described by the basic hydrodynamic equations, the continuity equation

$$\partial_t \rho + \nabla.(\rho u) = 0 \qquad (A\text{-}18)$$

which expresses the conservation of mass, and the momentum equation

$$\partial_t (\rho u) + \nabla.(\rho u) = -\nabla p + \nabla.\tau + \rho g \qquad (A\text{-}19)$$

which expresses the conservation of momentum. Here ρ is the fluid density, u the fluid velocity, p the hydrostatic pressure, τ the fluid stress tensor, and g the acceleration due to external forces including e.g. the effect of gravity on the fluid.

The equation for energy conservation can be written as

$$\rho \frac{d\hat{u}}{dt} + p(\nabla . u) = \nabla . (k \nabla T) + \Phi \qquad \text{(A-20)}$$

where T is temperature, k the coefficient of thermal conductivity of the fluid, Φ the viscous dissipation function, and the density of thermal energy $\hat{u} = \hat{u}(p, T)$ is often approximated such that $d\hat{u} \approx c_v dT$, where c_v is the specific heat.

At low Reynolds numbers, the most important relation describing fluid transport through porous media is Darcy's law

$$q = -\frac{k}{\mu} \nabla p \qquad \text{(A-21)}$$

where q is the volumetric fluid flow through the (homogeneous) medium and k is the permeability coefficient that measures the conductivity to fluid flow of the porous material.

TRANSPORT MECHANISMS IN POROUS MEDIA

The study of flow systems which compose of a porous medium and a homogenous fluid has attracted much attention since they occur in a wide range of the industrial and environmental applications. Examples of practical applications are: flow past porous scaffolds in bioreactors, drying process, electronic cooling, ceramic processing, and overland flow during rainfall, and ground-water pollution.

In the single-domain approach, the composite region is considered as a continuum and one set of general governing equations is applied for the whole domain. The explicit formulation of boundary conditions is avoided at the interface and the transitions of the properties between the fluid and porous medium are achieved by certain artifacts. Although this method is relatively easier to implement, the flow behavior at the interface may not be simulated properly, depending on how the code is structured.

In the two-domain approach, two sets of governing equations are applied to describe the flow in the two regions and additional boundary conditions are applied at the interface to close the two set of equations. This method is more reliable since it tries to simulate the flow behavior at the interface. Hence, in the present study, the two-domain approach, and the implementation of the interface boundary conditions, will be considered.

Fluid flow in a porous medium is a common phenomenon in nature, and in many fields of science and engineering. Important everyday flow phenomena include transport of water in living plants and trees, and fertilizers or wastes in soil. Moreover, there is a wide variety of technical processes that involve fluid dynamics in various branches of process industry. The importance of improving our understanding of such processes arises from the high amount of energy consumed by them. In oil recovery, for example, a typical problem is the amount of un-recovered oil left in oil reservoirs by traditional recovery techniques. In many cases the porous structure of the medium and the related fluid flow are very complex, and detailed studies of these flows pose demanding tasks even in the case of stationary single-fluid flow. In experimental and theoretical work on fluid flow in porous materials it is typically relevant to find correlations between material characteristics, such as porosity and specific surface area, and flow properties. The most important phenomenological law governing the flow properties, first discovered by Darcy, defines the permeability as conductivity to fluid flow of the porous material. Permeability is given by the coefficient of linear response of the fluid to a non-zero pressure gradient in terms of the flux induced.

Some of the material properties that affect the permeability, e.g. tortuosity, are difficult to determine accurately with experimental techniques, which have been, for a long time, the only practical way to study many fluid-dynamical problems. Improvement of computers and the subsequent development of methods of computational fluid dynamics (CFD) have gradually made it possible to directly solve many complex fluid-dynamical problems. Flow is determined by its velocity and pressure fields, and the CFD methods typically solve these in a discrete computational grid generated in the fluid phases of the system. Traditionally CFD has concentrated on finding solution to differential continuum equations that govern the fluid flow. The results of many conventional methods are sensitive to grid generation which most often can be the main effort in the application. A successfully generated grid is typically an irregular mesh including knotty details that follow the expected streamlines.

Transport in a porous media can be due to several different mechanisms. Three of these mechanisms are often considered most dominant: molecular diffusion, capillary diffusion, and convection (*Darcy flow*).

The *Darcy law* has been derived as follows: we consider a macroscopic porous medium which has a cross section A and overall length L, and we impose an oriented fluid flow rate $\vec{Q}$, to flow through it .When a steady state is reached, the induced hydrostatic pressure gradient $\vec{\nabla}p$ is related to $\vec{Q}$ by the vectorial formula:

Fluid dynamics (also called fluid mechanics) is the study of moving (deformable) matter, and includes liquids and gases, plasmas and, to some extent, plastic solids. From a 'fluid-mechanical' point of view, matter can, in a broad sense, be considered to consist of fluid and solid, in a one-fluid system the difference between these two states being that a solid can resist shear stress by a static deformation, but a fluid can not . Notice also that thermodynamically a distinction between the gas and liquid states of matter cannot be made if temperature is above that of the so-called critical point, and below that temperature the only essential differences between these two phases are their differing equilibrium densities and compressibility.

$$\frac{\vec{Q}}{A} = \frac{\overline{K}}{\mu_f} \cdot (\vec{\nabla}p - \rho_f \cdot \vec{g}) \Leftrightarrow \vec{v}_m = \frac{\overline{K}}{\mu_f} \cdot \left(\frac{\Delta p}{L} - \rho_f \cdot \vec{g} \right) \qquad (A\text{-}22)$$

where $\vec{g}$ is the acceleration of the gravity field, ρ_f and μ_f are respectively the *specific mass* and the *dynamic viscosity* of fluid, $\vec{v}_m$ the *filtration velocity* over the cross section A. Formula (A.22) defines a second order symmetrical tensor $\overline{K}$, the *permeability*. It takes into account the macroscopic influence of the porous structure from the "resistance to the flow" point of view. The more permeable a porous medium is, the less it will resist to an imposed flow. The *permeability* is an intrinsic property of the porous matrix, based only on geometrical considerations, and is expressed in $[m^2]$. The tensorial character of $\overline{K}$ reflects the porous matrix *anisotropy*.

At the surface of the textile, two processes occur simultaneously in drying: heat transfer from the air to the drying surface and mass transfer from the drying surface to the surrounding air. The energy transfer between a surface and a fluid moving over the surface is traditionally described by convection. The unbound moisture on the surface of the material is first vaporised during the constant drying rate period.

Heat transfer by convection is described as

$$\frac{dQ}{dt} = \bar{h}A(T_A - T_S) \qquad (A\text{-}23)$$

where dQ/dt is the rate of heat transfer, $h[W/m^2K]$ is the average heat transfer coefficient for the entire surface, A. T_S is the temperature of the material surface and T_A is the air temperature. The temperature on the surface is close to the wet bulb temperature of the air when unbound water is evaporated (Bejan et al.2004).

A similar equation describes the convective mass transfer. The total molar transfer rate of water vapour from a surface, dN_v/dt [kmol/s], is determined by

$$\frac{dN_v}{dt} = \bar{h}_m A(C_{v,A} - C_{v,S}) \qquad (A\text{-}24)$$

where $\bar{h}_m$ [m/s] is the average convection mass transfer coefficient for the entire surface, $C_{v,A}$ is the molar concentration of water vapour in the surrounding air and $C_{v,S}$ is the molar concentration on the surface of the solid with the units of [kmol/m3]. During the constant drying rate period the drying rate is controlled by the heat and/or mass transfer coefficients, the area exposed to the drying medium, and the difference in temperature and relative humidity between the drying air and the wet surface of the material (Bejan et al. 2004).

The average convection coefficients depend on the surface geometry of the material and the flow conditions. The heat transfer coefficient, h, can be determined by the average Nusselt number, $\overline{Nu}$:

$$\overline{Nu} = \frac{\bar{h}L}{k_A} = f(\text{Re}, \text{Pr}) \qquad (A\text{-}25)$$

where k_A is the heat conductivity for the air and L is the characteristic length of the surface of interest. $\overline{Nu}$ shows the ratio of the heat transfer that depends on convection to the heat transfer that depends on conduction in the boundary layer. The Nusselt number is a function of the Reynold number, Re, and the Prantdl- number, Pr. Pr is the relation between the thickness of the thermal and the velocity boundary layers. If $Pr=1$, the thickness of the thermal and velocity boundary layers are equal. For air $Pr=0.7$. To determine the mass transfer coefficient, $\overline{h}_m$, the average Sherwood-number, $\overline{Sh}$ is used:

$$\overline{Sh} = \frac{\overline{h}_m L}{D_{AS}} = f(\text{Re}, Sc) \tag{A-26}$$

where D_{AS} is the diffusion coefficient. $\overline{Sh}$ is a function of the Reynold number, Re, and the Schmidt number, Sc, which is the relation between the thickness of the concentration and the velocity boundary layers.

Water vapor in the porous media can move by molecular or Fickian diffusion if the pores are large enough. Molecular diffusion is described by Fick's law

$$J = -D \frac{\partial c}{\partial x} \tag{A-27}$$

D is the molecular diffusivity.

In a fiber such as textile the diffusion does not only depend on the difference in concentration but also on the characteristics of the textile. He describes the moisture movement as being dependent on the density of the solid, which is a function of the moisture content as the fiber swells or shrinks in response to the moisture that is present.

Flow in porous media plays an important role in many areas of science and engineering. Examples of the application of porous media flow phenomena are as diverse as flow in human lungs or flow due to solidification in the mushy zone of liquid metals.

Flow in porous media is difficult to be accurately modeled quantitatively. Richards equation can give good results, but needs constitutive relations. These are usually empirically based and require extensive calibration. The parameters needed in the calibration are amongst others: capillary pressure and pressure gradient, volumetric flow, liquid content, irreducible liquid content, and temperature. In practice it is usually too demanding to measure all these parameters.

The description of the behavior of fluids in porous media is based on knowledge gained in studying these fluids in pure form. Flow and transport phenomena are described analogous to the movement of pure fluids without the presence of a porous medium. The presence of a permeable solid influences these phenomena significantly. The individual description of the movement of the °uid phases and their interaction with the solid phase is modeled by an up-scaled porous media flow equation. The concept of up-scaling from small to large scales is widely used in physics. Statistical physics translates the description of individual molecules into a continuum description of different phases, which in turn is translated by volume averaging into a continuum porous medium description.

REFERENCES

Armour, J., and Cannon, J., Fluid Flow Through Woven Screens, *AIChE J.* 14, (3), 415-420 (1968).

ASTM D737-75, Standard Test Methods for Air Permeability of Textile Fabrics.

ASTM E96-95, Water Vapour Transmission of Materials.

Arnold, G. and Fohr, J.P., Slow Drying Simulation in Thick Layears of Granular Products, *Int. J. Heat Mass Transfer* 31 (12), 2517-2562 (1988).

Azizi, S., Moyne, C., and Degiovanni, A., Approche Experimentale et theorique de la Conductivite Thermique des Milieux Poreux Humides, *Int. J. Heat Mass Transfer* 31(11), 2305-2317 (1988).

Backer, S., The Relationship Between the Structural Geometry of a Textile Fabric and its Physical Properties, Part IV: Interstice Geometry and Air Permeability, *Textile Res. J.* 21, 703-714 (1951).

Barnes, J., and Holcombe, B., Moisture Sorption and Transport in Clothing During Wear, *Textile Res. J.* 66 (12), 777-786 (1996).

Bartles, V. T., Survey on the Moisture Transport Properties of Foul Weather Protective Textiles at Temperatures Around and Below the Freezing Point, Technical Report no. 11674, Hohenstein Institute of Clothing Physiology, Boennigheim, Germany, 2001.

Bears, J., *"Dynamics of Fluids in Porous Media"*, Elsevier, New York, 1972.

Black, W.Z. and Hartley, J.G., *Thermodynamics,* Harper & Row, New York, 1985.

BS 4407, *Quantitative Analysis of Fiber Mixtures,* 1997.

BS 7209, *Specification for Water Vapour Permeable Apparel Fabrics,* 1990.

CAN2-4.2-M77, *Method of Test for Resistance of Materials to Water Vapour Diffusion* (Control Dish Method), 1977.

CGSB-4.2 No. 49-M91, *Resistance of Materials to Water Vapour Diffusion.*

Chen, C. S. and Johnson, W. H., Kinetics of Moisture Movement in Hygroscopic Materials, In: Theoretical Considerations of Drying Phenomenon, *Trans. ASAE.,* 12, 109-113 (1969).

Chen, P. and Pei, D., A Mathematical Model of Drying Process, *Int. J. Heat Mass Transfer* 31 (12), 2517-2562 (1988).

Chen, P., Schmidt, P.S., *An Integral Model for Drying of Hygroscopic and Non-hygroscopic Materials with Dielectric Heating, Drying Technol.* 8 (5), 907-930 (1990).

Chen, P. and Schmidt, P.S., A Model for Drying of Flow-through Beds of Granular Products with Dielectric Heating, in: *Transport Phenomena in Materials Processing, American Society of Mechanical Engineers, Heat Transfer Division,* (Publication) HDT, vol. 146, ASME, New York, 121-127 (1990).

Davis, A., and James, D., Slow Flow Through a Model Fibrous Porous Medium, *Int. J. Multiphase Flow* 22, 969-989 (1996).

Dietl, C. and George, O. P., and Bansal, N. K. , *Modeling of Diffusion in Capillary Porous Materials During the Drying Process, Drying Technol.* 13 (1&2), 267-293 (1995).

Ea, J. Y., *Water Vapour Transfer in Breathable Fabrics for Clothing*, PhD thesis, University of Leeds, 1988.

Flory, P.J., *Statistical Mechanics of Chain Molecules,* Interscience Pub. NY, 1969.

Francis, N. D. and Wepfer, W. J. , Jet Impeingement Drying of a Moist Porous Solid, *Int. J. Heat Mass Transfer* 39 (9), 1911-1923 (1996).

Gerald, C. F. and Wheatley, Applied Numerical Analysis, Fourth ed., Addison-Weseley, Reading, MA, 1989.

Ghali, K., Jones, B., and Tracy, E., Experimental Techniques for Measuring Parameters Describing Wetting and Wicking in Fabrics, *Textile Res. J.,* 106-111 (1994).

Gibson, P., Elsaiid, A., Kendrick, C.E., Rivin, D., and Charmchi, M., A Test Method to Determine the relative Humidity Dependence of the Air Permeability of Textile Materials, *J. Testing Eval.* 25(4), 416-423 (1997).

Gibson, P. and Charmchi, M., The Use of Volume-Averaging Techniques to Predict Temperature Transients Due to Water Vapour Sorption in Hygroscopic Porous Polymer Materials, *J. Appl. Polym. Sci.* 64, 493-505 (1997).

Ghali, K., Jones, B., and Tracy, E., Modeling Heat and Mass Transfer in Fabrics, *Int. J. Heat Mass Transfer* 38(1), 13-21 (1995).

Gennes, P.G., Scaling Concepts in Polymer Physics, 3rd ed., Cornell University Press, Ithaca, NY, 1988.

Givoni, B., and Goldman, R.F., Predicting Metabolic Energy Cost, *J. Appl. Physiol.* 30(3), 429-433 (1971).

Green, J. H., "An Introduction to Human Physiology" J. Comyn, Ed., Elsevier Applied Science Publishers, London, 1985.

Greenkorn, R.A., "Flow Phenomena in Porous Media" Marcel Dekker, New York, 1984.

Hadley, G. R., Numerical Modeling of the Drying of Porous Materials, in: *Proceedings of The Fourth International Drying Symposium*, vol. 1, 151-158 (1984).

Haghi, A.K., Moisture permeation of clothing, *JTAC* 76, 1035-1055(2004).

Haghi, A.K., Thermal analysis of drying process, *JTAC* 74, 827-842(2003).

Haghi, A.K., Some Aspects of Microwave Drying , The Annals of Stefan cel Mare University, Year VII, No. 14, 22-25 (2000).

Haghi, A.K., A Thermal Imaging Technique for Measuring Transient Temperature Field- An Experimental Approach, The Annals of Stefan cel Mare University, Year VI, No. 12, 73-76 (2000).

Haghi, A.K., Experimental Investigations on Drying of Porous Media using Infrared Radiation , *Acta Polytechnica,* 41(1), 55-57 (2001).

Haghi, A.K., A Mathematical Model of the Drying Process , *Acta Polytechnica* 41(3) 20- 23 (2001).

Haghi, A.K., Simultaneous Moisture and Heat Transfer in Porous System, *Journal of Computational and Applied Mechanics* 2(2), 195-204 (2001).

Haghi, A.K., A Detailed Study on Moisture Sorption of Hygroscopic Fiber, *Journal of Theoretical and Applied Mechanics* 32(2) 47-62 (2002).

Haghi, A.K., A mathematical Approach for Evaluation of Surface Topography Parameters , *Acta Polytechnica* 42(2) 35- 40 (2002).

Haghi, A. K., Mechanism of Heat and Mass Transfer in Moist Porous Materials, *Journal of Technology,* 35(F).1-16 (2002).

Haghi, A.K., A Study of Drying Process , *H.J.I.C.,* 30, 261-269 (2002).

Haghi, A.K., Experimental Evaluation of the Microwave Drying of Natural Silk *J. of Theoretical and Applied Mechanics* 33, 83-94, (2003).

Haghi , A. K., Mahfouzi, K. and Mohammadi, K., The effects of Microwave Irradiations on Natural Silk , *JUCTM* 38, 85-96 (2002).

Haghi, A.K., The Diffusion of Heat and Moisture Through Textiles, *International Journal of Applied Mechanics and Engineering* 8(2), 233-243 (2003).

Haghi , A.K., and Rondot, D., Heat and Mass Transfer of Leather in The Drying Process, *IJC&Chem. Engng.* (23), 25-34 (2004).

Haghi, A.K. , Heat and Mass Transport Through Moist Porous Materials, *14th Int. Symp. on Transport Phenomena Proc.*, 209-214, 6-9 July 2003, Indonesia.

Hartley, J.G., *Coupled Heat and Moisture Transfer in Soils: A Review, Adv. Drying 4* , 199-248 (199-248).

Higdon, J., anf Ford, G., Permeability of Three-Dimensional Models of Fibrous Porous Media, *J. Fluid Mechan,* 308, 341-361 (1996).

Hong, K., Hollies, N. R. S., and Spivak, S. M., Dynamic Moisture Vapour Transfer Through Textiles, Part I: Clothing Hygrometry and the Influence of Fiber Type, *Textile Res. J.* 58(12), 697-706 (1988).

Hsieh, Y. L., Yu, B., and Hartzell, M., Liquid Wetting Transport and Retention Properties of Fibrous Assemblies, Part II: Water Wetting and Retention of 100% and Blended Woven Fabrics, *Textile Res. J.* 62(12), 697-704 (1992).

Huh, C. and Scriven, L. E., Hydrodynamic Model of Steady Movement of a Solid-Liquid-Fluid Contact Line, *J. Coll. Inter. Sci*, 35, 85-101 (1971).

Incropera, F. P., Dewitt, D.P., Fundamentals of Heat and Mass Transfer, second ed., Wiley, New York, 1985.

ISO 11092, Measurement of Thermal and Water-vapour Resistance under Steady-state Conditions (Sweating Guarded-hotplate Test), 1993.

Ito, H., and Muraoka, Y., Water Transport Along Textile Fibers as Measured by an Electrical Capacitance Technique, *Textile Res. J.* 63(7), 414-420 (1993).

Jackson, J., and James, D., The Permeability of Fibrous Porous Media, *Can. J. Chem. Eng.* 64, 364-374 (1986).

Jacquin, C. H. and Legait, B., Influence of Capillarity and Viscosity During Spontaneous Imbibition in Porous Media and Capillaries, *Phy.-Chem. Hydro.* 5, 307-319 (1984).

Jirsak, O., Gok, T., Ozipek, B., and Pau, N., Comparing Dynamic and Static Methods for Measuring Thermal Conductive Properties of Textiles, *Textile Res. J.* 68(1), 47-56 (1998).

Kaviany, M., "Principle of Heat Transfer in Porous Media", Springer, New York, 1991.

Keey, R.B., "Drying: Principles and Practice", Oxford, Pergamon, 1975.

Keey, R. B., "Introduction to Industrial Drying Operations", Oxford, Pergamon, 1978.

Keey, R.B., The Drying of Textiles, *Rev. Prog. Coloration* 23, 57-72 (1993).

Kulichenko, A., and Langenhove, L., The Resistance to Flow Transmission of Porous Materials, *J. Textile Inst.* 83 (1), 127-132 (1992)

Kyan, C., Wasan, D., and Kintner, R., Flow of Single-Phase Fluid through Fibrous Beds, *Ind. Eng. Chem. Fundament.* 9 (4), 596-603 (1970).

Le, C. V., and Ly, N. G., Heat and Moisture Transfer in Textile Assemblies, Part I: Steaming of Wool, Cotton, Nylon, and Polyester Fabric Beds, *Textile Res. J.* 65(4), 203-212 (1995).

Le, C. V., Tester, D. H., and Buckenham, P., Heat and Moisture Transfer In Textile Assemblies, Part II: Steaming of Blended and Interleaved Fabric-Wrapper Assemblies, *Textile Res. J.* 65(5), 265-272 (1995).

Lee, H. S., Carr, W. W., Becckham, H. W., and Wepfer, W.J., Factors Influencing the Air Flow Through Unbacked Tufted Carpet, *Textile Res. J.* 70, 876-885 (2000).

Lee, H. S., Study of the Industrial Through-Air Drying Process For Tufted Carpet, Doctoral thesis, Georgia Institute of Technology, Atalnta, GA, 2000.

Luikov, A. V., Heat and Mass Transfer in Capillary Porous Bodies, Pergamon Press, Oxford, 1966.

Luikov, A.V., Systems of Differential Equations of Heat and Mass Transfer in Capillary Porous Bodies, *Int. J. Heat Mass Transfer* 18, 1-14, 1975.

Metrax, A. C. and Meredith R. J., Industrial Microwave Heating, Peter Peregrinus Ltd, London, England, 1983.

Mitchell, D. R., Tao, Y. and Besant, R. W., Validation of Numerical Prediction for Heat Transfer with Airflow and Frosting in Fibrous Insulation, paper 94-WA/HT-10 in *"Proc. ASME Int. Mech. Eng. Cong., Chicago,"* Nov. 1994.

MOD Specification UK/SC/4778A SCRDE, Moisture Vapour Transmission Test Method.

Morton, W. E., and Hearle, J. W. S., Physical Properties of Textile Fibers" 3rd ed., Textile Institute, Manchester, U.K., 1993.

Moyene, C., Batsale, J.C., and Degiovanni, A., Approche Experimentale et theorique de la Conductivite Thermique des Milieux Poreux Humides, II: Theorie, *Int. J. Heat Mass Transfer* 31(11), 2319-2330 (1988).

Mujumdar, A. S., Handbook of Industrial Drying, Marcel Decker, New York, 1985.

Nasrallah, S.B. and Pere, P., Detailed Study of a Model of Heat and Mass Transfer During Convective Drying of Porous Media, *Int. J. Heat Mass Transfer* 31 (5), 957-967 (1988).

Nossar, M.S., Chaikin, M.,, and Datyner, A., High Intensity Drying of Textile Fibers, Part I: The Nature of The Flow of Air Through Beds of Drying Fibers, *J. Textile Inst,* 64, 594-600 (1973).

Patankar, S. V., "Numerical Heat Transfer and Fluid Flow", Hemisphere Publishing, NY, 1980.

Penner, S., and Robertson, A., Flow through Fabric-Like Structures, *Textile Res. J.* 21, 775-788 (1951).

Provornyi, S., and Slobodov, E., Hydrodynamics of a Porous Medium with Intricate Geometry, *Theoret. Foundat. Chem. Eng.* 29 (1), 1-5 (1995).

Rainard, L. W., Air Permeability of Fabrics I, *Textile Res. J.* 16, 473-480 (1946).

Rainard, L. W., Air Permeability of Fabrics II, *Textile Res. J.* 17, 167-170 (1947).

Renbourne, E.T., "Physiology and Hygiene of Materials and Clothing," Merrow Publishing Company, Herts, England, 1971.

Saltiel, C. and Datta, A.S., Heat and Mass Transfer in Microwave Processing, *Advances in Heat and Mass Transfer* 33, 1-94 (1999).

Sanga, E., Mujumdar, A. S., Raghavan, G. S., Heat and Mass Transfer in Non-homogeneous Materials under Microwave Field, Presented at The 50th Canadian Chemical Engineering Conference, Montreal, Canada, October 15-18, 2000.

Simacek, P., and Advani, S., Permeability Model for a Woven Fabric, *Polym. Compos.* 17 (6), 887-899 (1996).

Spencer-Smith, J.L., The physical Basis of Clothing Comfort, Part 3: Water Vapour Transfer Through Dry Clothing Assemblies, *Clothing Res. J.* 5, 82-100 (1977).

Spilman, L., and Goren, S., Model for Predicting Pressure Drop and Filtration Efficiency in Fibrous Media, *Environ. Sci. Techno.* 2, 279-287 (1968).

Stanish, M. A., Schajer, G.S., and Kayihan, F., A Mathematical Model of Drying for Hygroscopic Porous Media, *AIChE J.* 32(8), 1301-1311 (1986).

Toei, R., Drying Mechanism of Capillary Porous Bodies, *Adv. Drying 2,* 269-297 (1983).

Umbach, K. H., Investigation of Constructional Principles for Clothing Textiles Made of Synthetic Fibers Worn Next to the Skin with Good

Comfort Properties, Technical Report no. AiF 3653, Hohenstein Institute of Clothing Physiology, Boennigheim, Germany, 1977.

Umbach, K.H., Moisture Transport and Wear Comfort in Microfiber Fabrics, *Melliand Engl,* 74, E78-E80 (1993).

Von Hippel, A. R., "Dielectric Materials and Applications", MIT Press, Boston, 1954.

Waananen, K.M., Litchfield, J. B., Okos, M. R., Classification of Drying Models foe Porous Solids, *Drying Technol.* 11 (1), 1-40 (1993).

Watkins, D.A., and Slater, K., The Moisture Vapour Permeability of Textile Fabrics, *J. Textile Inst.* 72, 11-18 (1981).

Williams, A., "Industrial Drying", Gardner/Leonard Hill Publishers, London, England, 1971.

M. Ziabari, V. Mottaghitalab, S. T. McGovern and A. K. Haghi, *Chim. Phys. Lett.,* 25, 3071 (2008).

M. Ziabari, V. Mottaghitalab, S. T. McGovern and A. K. Haghi, *Nanoscale Research Letter,* 2, 297(2007).

M. Ziabari, V. Mottaghitalab and A. K. Haghi, *Korean J. Chem. Eng.,* 25, 919 (2008).

M. Ziabari, V. Mottaghitalab and A. K. Haghi, *Korean J. Chem. Eng.,* 25, 923 (2008).

M. Ziabari, V. Mottaghitalab and A. K. Haghi, *Korean J. Chem. Eng.,* 25, 905 (2008).

A. K. Haghi and M. Akbari, *Physica Status Solidi,* 204, 1830 (2007).

M.Kanafchian, M.Valizadeh and A .K.Haghi, *Korean J. Chem. Eng.,* 28, 428 (2011).

M.Kanafchian, M.Valizadeh and A .K.Haghi, *Korean J. Chem. Eng.,* 28, 763 (2011).

M.Kanafchian, M.Valizadeh and A .K.Haghi, *Korean J. Chem. Eng.,* 28, 751 (210 M.Kanafchian, M.Valizadeh and A .K.Haghi, *Korean J. Chem. Eng.,* 28, 445(2011).

A. Afzali, V. Mottaghitalab, M.Motlagh, A.K.Haghi, *Korean J. Chem. Eng.,* 27, 1145(2010).

Z.Moridi, V. Mottaghitalab, A. K. Haghi, *Korean J. Chem. Eng.,* 28, 445(2011).

A.K.Haghi, *Cellulose Chem. Technol.,* 44, 343 (2010)

Z.Moridi, V. Mottaghitalab, A. K. Haghi, *Cellulose Chem. Technol.,* 45, 549 (2011)

INDEX

A

acoustics, ix
activation energy, xi, 12, 32, 71, 85
AD, 107
adsorption, 12, 51, 52, 86, 90, 102, 105
adverse effects, 95
air temperature, xii, 37, 120, 140
algorithm, 101
alters, 24
ambient air, 19, 52
anisotropy, 139
apparel products, 3
assessment, 108
atmosphere, 11, 19, 69, 81, 86, 105
atoms, 7, 58, 61, 127

B

barriers, 112
base, 97
bending, 63
benefits, 57, 60, 64, 79
bleeding, 113, 121
blends, 111
bonds, 12, 25
bulk materials, 64

C

calibration, 141
capillary, xii, xiii, 3, 11, 26, 27, 28, 29, 30, 31, 42, 43, 89, 105, 108, 119, 121, 122, 129, 131, 138, 141
carbon, 21
ceramic, 137
chain molecules, 24
chemical, 3, 12, 16, 21, 49, 50, 120
chemical reactions, 21
Chicago, 146
clothing, 11, 81, 82, 83, 84, 85, 87, 96, 97, 98, 99, 100, 105, 107, 108, 110, 112, 144
coatings, 60, 84
collisions, 50
color, 7, 24
commercial, 17, 103
complexity, 87, 100
complications, 129
composites, vii, 79
composition, 22, 49, 52
compressibility, 139
compression, 74
computation, 101
computational fluid dynamics, 138
computational grid, 138
condensation, vii, 1, 3, 84, 90, 91, 103, 105, 108, 118, 120

conditioning, 82
conductance, 48, 51
conduction, vii, 1, 2, 3, 5, 9, 11, 17, 24, 45, 46, 48, 49, 50, 52, 55, 59, 68, 70, 81, 82, 84, 90, 113, 114, 118, 119, 120, 125, 140
conductivity, xi, xii, 6, 21, 29, 30, 31, 32, 35, 36, 48, 49, 50, 64, 66, 70, 71, 81, 90, 91, 92, 102, 116, 117, 120, 124, 125, 137, 138
conductor, 58
conductors, 127
conference, 107
configuration, 60
conservation, 35, 70, 84, 86, 136
constant rate, 15, 16, 17, 18, 23, 27, 48, 78
construction, 24, 90
consumers, 90
cooling, 4, 24, 46, 81, 85, 122, 137
correlation, 21, 30, 88
correlations, 21, 138
cost, 15, 19, 59
cotton, vii, 20, 83, 85, 86, 88, 89, 90, 91, 95, 96, 97
covalent bond, 7, 58, 127
creep, 12
crises, 59
critical value, 27
crystallites, 25
crystallization, 25

D

DEA, ix
deformation, 139
depth, 66, 68, 70
desorption, xii, 84, 120
dielectric constant, 65, 66, 67
dielectrics, 7, 58, 66, 127
differential equations, 83, 84, 86
diffusion, vii, xi, xii, 1, 2, 3, 10, 11, 12, 18, 24, 31, 32, 33, 34, 36, 38, 43, 50, 51, 52, 71, 81, 84, 85, 86, 87, 89, 100, 101, 102, 108, 109, 110, 111, 112, 118, 119, 122, 130, 134, 138, 141
diffusion process, 31

diffusivity, xi, 70, 71, 76, 141
discomfort, 83, 85, 88, 95
dispersion, 9
distillation, 9, 83
distribution, 1, 3, 9, 10, 23, 36, 38, 57, 66, 120, 130, 131, 132, 133
drawing, 1
drying, 2, 3, 4, 7, 8, 11, 15, 16, 17, 18, 19, 20, 22, 23, 25, 26, 27, 28, 31, 32, 39, 41, 42, 43, 58, 59, 60, 72, 73, 74, 76, 78, 79, 80, 97, 111, 114, 118, 119, 120, 121, 122, 127, 128, 135, 137, 139, 140, 144
drying gas, 20, 23
dyeing, 24, 60
dyes, 24, 60
dynamic viscosity, 139

E

effusion, 122
electric field, 7, 57, 58, 66, 69, 127
electrical conductivity, 48
electromagnetic, vii, 3, 6, 7, 8, 57, 58, 59, 64, 65, 68, 69, 113, 116, 118, 119, 126, 127, 128, 129
electromagnetic energy transmission, 3, 6, 57, 119, 126
electromagnetic spectrum, 6, 57, 64, 65, 126
electromagnetic waves, vii, 68, 113, 116, 118, 129
electron, 7, 58, 63, 127
electrons, 6, 57, 61, 63, 126
emitters, 60
energy, vii, 2, 3, 5, 6, 7, 8, 15, 16, 24, 25, 27, 35, 46, 50, 57, 58, 59, 60, 61, 63, 64, 65, 67, 68, 70, 79, 82, 86, 97, 103, 113, 116, 117, 118, 119, 125, 126, 127, 128, 129, 136, 138, 139
energy conservation, 136
energy consumption, 8, 59, 128
energy transfer, 24, 67, 116, 128, 139
engineering, ix, 3, 45, 79, 138, 141
England, 107, 146, 147, 148
entropy, 47, 48

environment, 11, 20, 22, 69, 74, 76, 81, 82, 85, 135
environmental conditions, 23, 88, 96
environments, 2, 11, 81, 94, 99
equality, 48
equilibrium, 12, 19, 23, 24, 32, 35, 54, 76, 84, 86, 87, 94, 98, 99, 100, 110, 130, 139
equipment, 15, 59, 128
evaporation, xi, xii, 2, 3, 11, 16, 17, 18, 19, 21, 22, 26, 27, 36, 37, 38, 39, 42, 72, 76, 82, 84, 85, 89, 90, 95, 97, 98, 103, 108, 118, 119
everyday life, 130
evolution, vii
excitation, 134
exercise, 83, 96, 98
experimental condition, 73
extreme cold, 99

F

far right, 17
fast processes, 89
feelings, 85
ferrite, 68
fertilizers, 138
fiber, xi, 1, 3, 12, 24, 37, 71, 86, 87, 89, 90, 91, 97, 99, 100, 101, 102, 105, 106, 111, 119, 121, 141
fibers, vii, xi, 1, 3, 12, 24, 25, 36, 37, 74, 81, 84, 85, 86, 87, 88, 89, 90, 91, 94, 95, 96, 97, 99, 100, 101, 102, 103, 112, 119, 121, 135
filament, vii, 25, 99
film thickness, 11
filters, 130
filtration, 139
Fisherman, 111
fluid, vii, 3, 4, 5, 9, 10, 18, 19, 21, 29, 33, 60, 103, 113, 115, 116, 118, 119, 122, 123, 124, 125, 131, 133, 134, 136, 137, 138, 139
fluid phase, 131, 138
food, 7, 43, 57, 58, 127
food industry, 43

footwear, 81
force, 12, 28, 37, 61, 62, 63, 89
Ford, 145
formation, 8, 58, 128
formula, 139
France, ix
freedom, 25
freezing, 43
friction, 8, 58, 127

G

gamma rays, 116, 128
gas diffusion, 84
geometry, 1, 8, 12, 48, 58, 59, 86, 115, 124, 128, 131, 140
Georgia, 146
Germany, 142, 148
glass transition, 24
glass transition temperature, 24
gravitational constant, 34, 75
gravitational effect, 33
gravity, 28, 60, 83, 131, 136, 139
grid generation, 138

H

heat capacity, 24, 35, 70, 71, 102
heat conductivity, 140
heat loss, 82, 96, 97, 98
heat release, 86, 89, 90
heat transfer, vii, xii, 2, 3, 4, 5, 6, 8, 9, 15, 17, 18, 19, 20, 21, 36, 37, 45, 46, 47, 48, 49, 53, 59, 67, 89, 90, 91, 98, 110, 113, 115, 116, 117, 118, 119, 120, 122, 123, 124, 125, 128, 129, 139, 140
heating rate, 8, 59, 128
history, 101
host, 60
hot spots, 59, 128
human, 94, 141
human body, 94
humidity, xiii, 4, 11, 12, 15, 16, 18, 19, 20, 23, 27, 32, 37, 75, 76, 81, 84, 85, 86, 87,

88, 89, 95, 96, 97, 101, 102, 106, 121, 140
hydrophilic materials, 12, 85
hydrophobic nature, 94
hydrophobic surface, 95
hyperthermia, 82
hypothermia, 81, 96

I

ideal, 6, 22, 35, 71, 126
immersion, 81
independent variable, 77
Indonesia, 145
industry, vii, 59, 138
inequality, 48
insulation, 87, 96, 97, 108, 111
insulators, 7, 58, 64, 127
interface, vii, 1, 45, 83, 118, 131, 137
inversion, 18, 20, 22, 23, 42, 69
ionic conduction, 67
Iran, ix
irradiation, 67
isotherms, 130
issues, 11, 81

J

Japan, 42

K

kinetics, 73, 87, 100, 105, 112

L

laminar, 28, 41, 115, 124
laws, 9, 45, 116, 128
lead, 8, 24, 57, 59, 96, 99, 101, 128, 130
leakage, 68
Lebanon, 103
light, 24, 57, 64, 82, 90, 116, 128
linear dependence, 8, 59, 128
linear function, 30, 49, 86

liquid droplet, 95
liquid phase, xiii, 33, 34, 103, 132
liquids, 59, 134, 139
Luo, 87, 100, 101, 110

M

machinery, 64
magnet, 135
magnetic field, 63, 134, 135, 136
magnetic moment, 134
magnitude, 50, 52, 61, 62, 63, 85, 135
majority, 133
man, 97, 98, 99
manufacturing, 2, 15, 60
mass, xi, xii, 1, 2, 17, 18, 19, 20, 21, 22, 25, 26, 27, 29, 33, 35, 37, 42, 43, 48, 50, 51, 52, 53, 61, 70, 74, 75, 84, 85, 86, 90, 91, 103, 108, 113, 118, 121, 136, 139, 140, 141
material porosity, 23
material surface, 3, 17, 59, 121, 140
materials, viii, ix, 3, 7, 8, 11, 19, 25, 26, 31, 43, 49, 57, 58, 59, 63, 64, 67, 68, 69, 70, 81, 84, 96, 108, 120, 121, 127, 128, 129, 130, 133
matrix, 1, 7, 9, 58, 94, 95, 127, 130, 131, 132, 133, 135, 139
matter, 61, 132, 139
measurement(s), 41, 50, 67, 83, 90, 99, 133, 134
mechanical stress, 25
media, 28, 42, 84, 109, 111, 119, 132, 135, 141, 142
medicine, vii
membrane permeability, 43
membranes, 31, 84
Metabolic, 144
metals, 63, 141
meter, 82
microclimate, 81
microstructure, 79
microwave heating, 7, 8, 58, 59, 60, 65, 67, 68, 69, 72, 73, 79, 127, 128
microwave radiation, 7, 58, 65, 127

Index

microwaves, 7, 57, 58, 59, 63, 64, 68, 70, 72, 78, 127
migration, 24, 60, 67, 72, 84, 108
mixing, 10
models, 87, 100, 103
moisture content, xi, xii, 2, 3, 8, 11, 12, 16, 17, 18, 19, 20, 23, 26, 27, 28, 29, 31, 36, 37, 39, 40, 58, 59, 66, 70, 71, 72, 73, 83, 86, 87, 98, 100, 103, 118, 121, 128, 129, 141
moisture sorption, 3, 87, 89, 100, 101, 119
mole, 21, 52
molecular structure, 45
molecular weight, xii, 32
molecules, 3, 6, 7, 8, 12, 21, 24, 45, 50, 57, 58, 119, 126, 127, 128, 134, 142
momentum, 103, 104, 136
MRI, 134, 135

N

nanofibers, vii
nanomaterials, vii
nanotechnology, vii
neglect, 45
neutral, 61
neutrons, 61
Newtonian wetting liquid, 132
NMR, 133, 134, 135
non-polar, 7, 58, 127
Nuclear Magnetic Resonance, 133, 134
nuclei, 61, 133
nucleus, 61, 134, 135

O

oil, 130, 138
one dimension, 77
operations, 19, 59
opportunities, 57
osmotic pressure, 122

P

parallel, 21
partial differential equations, 76
pathways, 3, 120
percolation, 27
percolation theory, 27
permeability, xii, 28, 29, 30, 31, 33, 34, 68, 74, 75, 81, 82, 84, 85, 99, 103, 131, 137, 138, 139
permeation, 85, 88, 96, 144
permittivity, 65, 66, 68, 69
physical activity, 81, 82
physical exercise, 85
physical mechanisms, 87
physical properties, 12, 25, 87, 102, 115, 116, 124
physical structure, 49
physics, 60, 142
physiological factors, 85
plants, 138
plasticization, 12
plastics, 66
polarization, 66
pollution, 137
polymer, viii, 24, 84, 85, 129
polymer chain, 24
polymer molecule, 85
polymers, 70
polypropylene, 87, 88, 90, 91, 96
polyurethane, 83
pore-size distribution (PSD), 133
porosity, 9, 29, 36, 74, 94, 103, 129, 138
porous materials, 121, 129, 130, 134, 138
porous media, 1, 33, 42, 43, 83, 103, 108, 118, 120, 121, 130, 131, 132, 135, 137, 138, 141, 142
pressure gradient, 11, 28, 73, 122, 138, 139, 141
principles, 20, 43, 60
probe, 67
proportionality, 48
protons, 61
PTFE, 83
publishing, 107

PVC, 83

R

radiation, vii, 2, 3, 6, 7, 8, 9, 11, 17, 18, 21, 46, 52, 57, 58, 63, 66, 69, 81, 82, 90, 113, 114, 116, 117, 118, 119, 120, 126, 127, 128, 129
Radiation, 3, 6, 21, 57, 116, 117, 119, 126, 144
radio, 57, 64
radius, 29
rainfall, 137
reactions, 57
reasoning, 12, 21
recovery, 138
regression, 73
relaxation, 8, 58, 128, 133, 134
relaxation times, 8, 58, 128
remediation, 8, 59, 128
repellent, 1, 85
requirements, 1
resilience, 25
resistance, 1, 11, 25, 81, 83, 84, 85, 90, 106, 115, 124, 131, 139
response, 138, 141
rubber, 7, 58, 83, 127
rubber treating, 7, 58, 127

S

saturation, xii, 16, 20, 23, 26, 29, 30, 37, 38, 39, 40, 74, 78, 91, 105, 130
savings, 8, 59, 128
scaling, 108, 142
science, 138, 141
scope, viii
selectivity, 69
sensation, 85, 95
sensations, 83, 87, 97
sensitivity, 134
shade, 60
shape, 1, 9, 25, 131
shear, 139
showing, 16

silkworm, 42
simulation, 87, 110
skin, 2, 11, 81, 83, 85, 88, 95, 96, 98, 99, 105, 106
society, 79
solid matrix, 1, 9, 130, 131, 132
solid phase, xii, 131, 132, 142
solidification, 141
solution, 54, 86, 87, 100, 101, 138
sorption, 3, 27, 31, 32, 71, 76, 84, 86, 87, 89, 91, 100, 101, 102, 106, 109, 112, 119
sorption kinetics, 86, 87, 100
sorption process, 86, 89, 100
species, xi, 21, 50, 51, 52, 67
specific heat, xi, 20, 24, 35, 36, 53, 137
specific surface, 138
spectroscopy, 133
spin, 133, 134
stability, 24, 25
state, 1, 22, 25, 27, 46, 48, 49, 51, 52, 84, 85, 86, 88, 90, 95, 114, 117, 125, 139, 145
states, 22, 46, 84, 89, 90, 130, 135, 139
steel, 130
stress, 136, 139
stretching, 24
structural relaxation, 24
structure, 1, 24, 29, 46, 121, 122, 132, 138, 139
supernatant pressure, 4, 121
surface area, 4, 18, 26, 122, 131
surface tension, xii, 3, 30, 89, 120
surplus, 82
survival, 89
susceptibility, 135, 136
sweat, 81, 82, 85, 88, 94, 95, 96, 98, 99
swelling, 25, 84
synthetic fiber, 24, 96, 97, 111

T

taffeta, 24
target, 17
techniques, 8, 64, 69, 84, 103, 128, 138
technology, 57, 59, 79

tension, 24, 25, 28
testing, 89, 105
textiles, 2, 3, 4, 15, 16, 24, 45, 81, 84, 85, 90, 105, 107, 109, 111, 114, 119, 121, 135, 136
texture, 7
thermal analysis, 1
thermal energy, 15, 50, 85, 95, 116, 129, 137
thermal properties, 8, 59, 128
thermal resistance, 87, 90, 97
thermodynamics, 2, 43, 45, 46, 48, 119
topology, 131
total energy, 46
transfer performance, 5, 123
transmission, vii, 3, 6, 50, 57, 68, 81, 83, 85, 95, 106, 107, 109, 111, 112, 113, 118, 119, 126
transport, 1, 3, 11, 16, 21, 31, 43, 80, 84, 85, 87, 88, 89, 90, 91, 94, 96, 100, 101, 103, 104, 109, 111, 113, 119, 121, 122, 137, 138, 142
transport processes, 80, 89
transportation, vii
treatment, 69, 79, 84
trial, 111

U

UK, 109, 146
uniform, 8, 51, 52, 58, 64, 66, 101, 128, 130
universal gas constant, 22
upholstery, 81
urban, ix
USA, ix
USDA, 107

V

vacuum, 6, 57, 116, 117, 126, 128

vapor, vii, xi, xiii, 1, 2, 3, 11, 12, 18, 19, 22, 23, 27, 31, 32, 34, 41, 42, 81, 82, 83, 84, 86, 87, 88, 89, 91, 94, 95, 96, 98, 99, 103, 105, 109, 118, 119, 121, 122, 133, 141
variables, 4, 6, 9, 16, 69, 86, 122, 125
variations, 24, 39, 59, 60, 69
velocity, xii, 5, 9, 11, 25, 29, 50, 64, 75, 115, 121, 123, 124, 136, 138, 139, 141
viscosity, 28, 33, 34
vision, 67
void space, xi, 3, 87, 89, 119, 129, 130, 132

W

walking, 82
Washington, 43, 107
waste, 128
water absorption, 97
water diffusion, 89
water evaporation, 41, 42
water sorption, 86, 89
water vapor, vii, 1, 3, 11, 12, 20, 32, 34, 71, 81, 82, 83, 84, 85, 86, 88, 89, 90, 94, 95, 96, 99, 100, 102, 105, 108, 114, 118, 120, 121, 136
wear, 11, 88, 91, 96
wetting, 1, 23, 84, 94, 132
wood, 7, 42, 43, 58, 127
wool, 19, 71, 84, 86, 87, 88, 90, 91, 92, 95, 96, 97, 100, 101, 102, 105, 106, 109, 110, 112
wool fibers, 86, 87, 95, 100, 112
World War I, 7, 58, 127
worldwide, 59

Y

yarn, 64, 99